Jill McWilliam's Book of

A Falcon Book published by Woodhead-Faulkner Ltd
in association with Bejam Group Ltd

Woodhead-Faulkner Ltd
7 Rose Crescent
Cambridge CB2 3LL

First published 1974
Fourth Impression, May 1976.

ISBN 0 85941 010 2

Printed in England by Sir Joseph Causton & Sons Limited

Foreword

by A. W. Perry, *Director, Bejam Group Ltd*

Few companies can have learned more about their business from their customers than Bejam has since we began in 1968.

In our stores we try to confine ourselves to facts. It still surprises many to be told not to buy this and to hear of disadvantages as well as the good news. However, this sort of frankness has helped bring about a rapport with customers which is both enduring and unusual.

The home economists in our stores, still known as Super Girls, are daily explaining the mysteries of freezing to newcomers and exchanging ideas with the more experienced freezer owners.

Increasingly I have felt the need for us to summarise this welter of information. Certainly there have been many excellent books in recent years on freezing, but there remained a gap between the questions that we are constantly asked and the answers contained.

We were fortunate in having Jill McWilliam to take on this task. Jill has grown up with home freezing during its formative years. She has acquired considerable knowledge of her subject whilst remaining in touch with many of our customers.

This is not a book about Bejam. It will not sell more peas or chops for us but if it helps strip freezing of some of its mystery then it will have performed a most useful service.

Jill McWilliam's Book of Freezing should prove invaluable to those about to buy their first freezer and worthwhile reading to those on their third.

Acknowledgements

The photographs on the following pages are reproduced with the kind permission of:
Alcan Polyfoil Limited: pages 53, 73, 76, 121
Alreston Kitchens: page 61
Birds Eye Foods Limited: page 65
Danish Food Centre, London: pages 80, 89
Dan-Maid Seafoods: pages 69, 122
W. & J. B. Eastwood Limited: page 57
Electrolux Limited: pages 15, 36, back cover
Flour Advisory Bureau: page 124
Jus-rol Limited: page 85
KPS Freezers Limited: pages 9, 13
Lakeland Plastics Limited: pages 38, 111
Lec Refrigeration Limited: pages 10, 91
Ross Poultry Limited: pages 56, 59
Suttons Seeds Limited: pages 100, 102, 104, 117, 119
Tupperware: page 110
Young's Seafoods Limited: page 78

Acknowledgements are also due to:
Deer Publishers Limited
Elsa Mayo for the front cover photograph

Contents

AUTHOR'S NOTE

All prices mentioned in this book were correct at the time of going to press, but obviously the prices of food and freezers are continually fluctuating. While always subject to change, the prices given will, however, serve as a useful guide to comparative costs.

Part One
Introducing the Freezer

1. WHY HAVE A DEEP FREEZER?

That's a very good question; you've probably managed very well for years without one and the fact that Mrs Jones down the road bought one last week doesn't seem reason enough to take the plunge. So let us analyse the advantages of having a freezer.

Firstly you can save money, and there are few who wouldn't consider that to be a good idea. Buying in larger quantities together with a certain amount of home freezing are the first things that spring to mind, but perhaps it is the less obvious points which are of major importance—anticipating price rises and buying accordingly, less shopping, saving petrol or bus fares, wastage cut to the minimum, to mention only a few. The old saying "Look after the pennies and the pounds will take care of themselves" still holds true today.

Unfortunately, a freezer doesn't come with a guarantee to save money—that's down to you—but many freezer owners save an average of 10 per cent on their food bills and indeed it is possible to save more by careful planning, common-sense buying and sensible use of the freezer. However, it is always difficult to estimate where saving money ends and enjoying better food for the same money begins. It is almost inevitable that your standard of living will rise, but to what extent is determined by you. If a higher standard of living can be achieved on equal food expenditure, to many that is preferable to a cash saving.

But the investment in a freezer can't be accounted for in purely money terms; the convenience it brings to the housewife is often the greatest advantage. A freezer is a shop of your own stocked with a wide variety of food to cater for all situations—illness in the family, unexpected guests, holidays, as well as the normal day-to-day requirements, such as packed lunches, quick snacks and ready-prepared meals.

Food shopping is cut to a minimum, and for many this is the major convenience. Entertaining is made simple, particularly at the festive season when the food can be prepared weeks in advance with no more last-minute panic.

A freezer is still considered a luxury item in Britain just as the refrigerator was 20 years ago. It is a piece of electrical equipment which is providing a means of saving money and offering convenience to the busy housewife. So you still may be saying, "But I have managed without one for years." Then spoil yourself and buy a machine which is going to be fun to use and give you a great deal of satisfaction.

Of course once you have a freezer it's difficult to know how you ever managed without it, but I would never underestimate the fun that can be obtained by using it, and often the sheer satisfaction of finding a bargain is worth more than the 5p saving—although these 5p's can transform themselves into a new hat in a very short space of time. Freezing the garden produce, experimenting with new foods and creating interesting recipes are not always economical and are time-consuming rather than time-saving, but a freezer can transform the monotony of family eating into a creative endeavour not only for the housewife but for all the family.

The advantages of a freezer are numerous and the secret is to let your freezer work for you—don't become its slave. With a little practice and a lot of common sense you can have a slice of economy, convenience and satisfaction—in fact there are very few electrical appliances which promise so much.

2. WHICH ONE AND WHERE?

There are various types of freezer on the market and it is easy to get bogged down with technical data. Firstly decide on the size of freezer required and its eventual location—one really has to consider the two together as in many homes space is a limiting factor.

The best advice to offer on size is: Buy the largest freezer you can accommodate. The repeated cry of many freezer owners is, "If only I'd bought a larger one." Although freezers may look overpowering in the showroom, do not underestimate how much you will want to store. Wander round the Food Hall in the Freezer Centre, examine the packs of food and remember you're buying for a couple of months and not for a week. You want to be able to store variety as well as price-saving bulk packs. The most popular size of freezer for a small family is 12 to 14 cu ft.

Location can be a problem. The kitchen is the obvious place,

A modern kitchen planned with freezers in mind

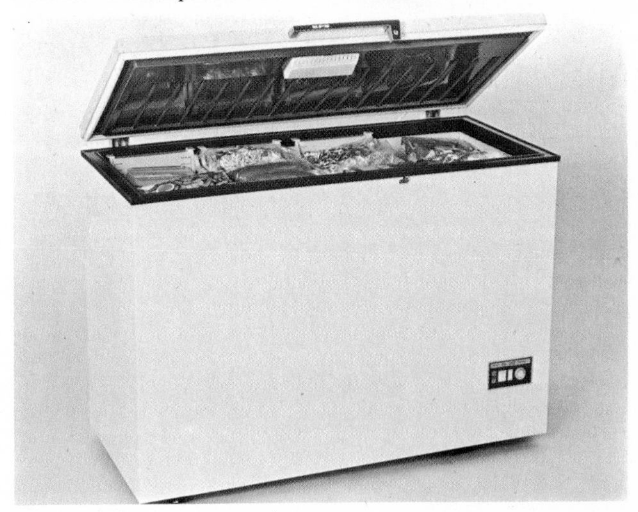
A 14.1 cu ft chest freezer

but space here is often restricted and other equipment takes priority. Without a doubt the freezer is most convenient if placed in the kitchen, but running costs can be 2p to 3p more per week because of other appliances creating a higher temperature.

A pantry or laundry room off the kitchen offers an ideal spot for the freezer if available, but, as few of us have the ideal spot, let us consider the alternatives.

Placing the freezer in an outhouse, garage or shed keeps running costs down to a minimum, but beware of damp. Condensation will cause any metal appliance to rust in time, but it can be halted if a few precautions are taken. Place the freezer on a plinth elevating it a couple of inches from the ground to ensure adequate ventilation. Coat with wax car polish periodically for added protection. Normal condensation will not affect the motor unit of the freezer in any way. The cabinet will deteriorate in time, but the working life of the freezer should not be affected.

A cellar often poses the same problems as an outbuilding, but an additional word of warning: if the cellar seems an ideal spot, measure the width of the freezer and check with the width of the stairway before ordering!

A freezer upstairs causes few difficulties. I would suggest a chest rather than an upright, so as not to cause weight problems, and it is wise to stand the freezer across the joists to distribute the weight more evenly. Choose a quiet-running machine so as not to disturb the light sleepers.

A conservatory by its very nature gets hot in the summer and thus wouldn't appear to be a suitable spot. Fan-assisted freezers, however, are designed to cope with high ambient temperatures, and although the running cost will be higher during a hot spell the freezer will function perfectly well, but allow air to flow around

Why not place the freezer in the garage?

the machine by means of an open window. If the freezer could be moved to another situation for the duration of a rare heatwave, so much the better.

The freezer can be sited virtually anywhere, but remember you will want to visit it at least once a day and so choose your location with that in mind.

Chest or Upright?

Which shape you choose will probably be determined by the location, and both will do an efficient job if sited correctly. But there are advantages and disadvantages for each which should be considered.

	Chest	Upright
Cost	12–14 cu ft £90–£110	12–14 cu ft £120–£140
Storage	20 lb–25 lb per cu ft	12 lb–15 lb per cu ft
Use	Slightly difficult to keep good stock rotation, requires organisation	Easy access and stock rotation. Sliding basket and door space helpful
Defrosting	Relatively easy, needs to be done 1–2 times yearly	Rather messy, needs to be done 3–4 times yearly. Automatic defrost available, but obviously more expensive
Space	Large amount of floor space required	Small amount of floor space required
Running Cost	In an ambient temperature 15°–18°C (60°–65°F) normal use, 20–22½p per week	In an ambient temperature 15°–18°C (60°–65°F) normal use, 22½–25p per week
Work Surface	Laminated work surface often fitted	Not applicable

Perhaps the upright freezers look more attractive and for the kitchen this must be taken into consideration. They are also easier to use and thus more convenient. However, if analysing economics, the chest would win. The choice is yours.

Fan, Skin or Static?

Not only do they come in different shapes and sizes but they function differently, too.

The different methods of refrigeration are merely a variation of the way the heat is extracted from the cabinet. Each performs the task of refrigeration satisfactorily, but their individual peculiarities could have a bearing on your choice.

	Fan-assisted condenser	**Skin-type condenser**	**Static plate condenser**
High temperatures 32°–38°C (90°–100°F)	Copes well with high temperatures for long periods	Copes with high temperatures for only a short time	Copes with high temperatures for only a short time
Home freezing capacity*	High 10–20% of maximum loading capacity	Average 10% of maximum loading capacity	Above average 10–15% of maximum loading capacity
Noise	A little noisy	Quiet	Average
Prone to condensation	Yes	No	A little
Home maintenance	Fan cleaned every six months	None	Condenser grilles to be kept free of dust
Surrounding space necessary	2–4 in. especially important at grille end	2–4 in. all round	2–4 in. all round

*It is difficult to generalise here as all freezers will perform differently. However, all machines will freeze down 10 per cent of their overall loading capacity. Some fan-assisted and static machines will do more, but it would rarely be a determining factor when buying unless vast quantities of home freezing were to be done.

Choose the machine which is best suited to your own family needs considering the four main factors:

Location.
Size.
Shape.
Performance.

Remember it's going to last you for the next 10 to 15 years and so you might as well choose the right one. Go to a reputable dealer who has a good range of freezers to offer, employs knowledgeable staff and preferably also sells the food, as this could influence your choice considerably.

A combined fridge/freezer—ideal for the small kitchen

3. IS INSURANCE REALLY NECESSARY?

As with most electrical appliances a freezer comes with a guarantee from the manufacturers stating that they accept responsibility for the machine during the warranty period. All reputable manufacturers can be relied upon to honour the guarantee and if the freezer breaks down during this period it will be repaired free of charge.

The problem, however, is one of time. If the machine is not repaired within 24 hours then the contents are at risk, and although the manufacturers will aim to be with you at their earliest convenience rarely would it be the same day. Their concern is for your freezer; the contents are not their responsibility.

The cost of the food stored in the freezer could be worth up to £100, and it is only common sense to safeguard against its loss. Many insurance companies will cover food at varying costs, and so shop around. Reasonable terms can be negotiated on household insurance policies, too, but read the small print first and ensure that claims are paid within a reasonable time limit.

But is insurance really solving the problem? Certainly it is essential, but it is cure rather than prevention. It is surely better to ensure that the machine can be repaired on an emergency basis, thus saving the food. Often home-grown produce and prepared dishes can't be assessed in money terms—a work of labour and love can't be compensated.

Maintenance contracts are not easy to come by and are quite expensive. Apart from the annual premiums being high, extra charges are often made for travelling and spare parts. Ensure that an emergency service operates daily before signing the contract.

Separate contracts for emergency repair and insurance will probably cost around £12 to £15 per year, which is a large slice out of the annual saving the freezer is making.

Some freezer food companies, however, operate their own schemes which are more reliable and better value for money. The schemes vary but in general they offer an emergency repair service and food insurance for an annual premium of under £12. After having sold you a freezer, it is in their own interests to keep it operating, thus keeping you, the food customer, happy. As they are responsible for insurance claims, their emergency service by necessity operates well. They couldn't afford to run the operation on any other basis. Some schemes insist on breakdown notification within 24 hours, which is valid in normal circumstances, but if you're on holiday this isn't always possible—so find a scheme which covers this eventuality.

As one would expect, you have to buy the freezer from the company to take advantage of these schemes, but ensure that

An upright freezer situated in an outhouse

you are not obliged to buy food also, as this is obviously very limiting.

Cover of some sort is essential. Freezers don't break down very often, but, if they do, it can involve a lot of expense. The best schemes are undoubtedly those which can be purchased with the freezer, but never sign anything before you've read the small print and enter into a contract only with a reputable company.

4. INSTALLATION

One talks about installing freezers, but in simple terms all that is required is a 13-amp plug and socket.

If on delivery the freezer has been up-ended in order to get it into its correct location, the oil in the compressor will have run into the pipes. Therefore, leave the machine to settle for an hour before switching on. This is not necessary if the machine has been kept comparatively level during transit.

Wash out the interior of the machine with a mild solution of bicarbonate of soda, and dry well. Switch on the machine and check that the lights are working.

The machine will probably be down to temperature within a matter of hours, but as a precaution allow it to run for 24 hours before filling with food, to ensure that it is functioning correctly. The walls should be covered with a crisp frost.

Few manufacturers provide thermometers, but they are a worthwhile investment. They can be acquired for less than £2 and tell you at a glance if the freezer is running normally. It is not always easy to assess an accurate thermostat setting, and it is essential that the freezer is run at –18° C (0°F) or just below. A higher temperature will affect the storage life of the food, but a freezer run at excessively low temperatures increases running costs unnecessarily.

When you have checked the temperature after 24 hours, the freezer can then be stocked with food.

5. HOME CARE

Fortunately, freezers need very little attention, but daily checking is necessary. A thermometer is an accurate guide, but the food itself will tell you if things are not right. Ice-cream is very sensitive to temperature: if it feels soft for no apparent reason it could mean that the cabinet temperature is too high. A warning light on the control panel is an early indication of a rise in temperature and so familiarise yourself with these warning signs and act immediately.

Before calling the engineer, do one or two checks yourself.

1. Has the fuse blown?
2. Has the introduction of too much "warm food" caused a temporary rise in temperature?
3. Was the cabinet door left ajar by mistake?
4. Is the fan unit or condenser grille clogged with dust?
5. Is the thermostat setting too high?

Once satisfied that a minor adjustment can't put things right, call the engineer immediately.

Home Maintenance

The action of the fan pulls cool air over the compressor, and in doing so attracts dust and fluff. It should be cleaned every six months or so to keep it in good working order.

1. Switch off the freezer at the mains.
2. Unscrew the inlet grille.
3. Remove the dust by careful use of a vacuum cleaner hose or soft brush.
4. Replace the grille and reconnect the freezer.

Static condensers need dusting at regular intervals.

Defrosting

It is not as daunting as it may at first appear. It need be done only three or four times a year in the case of an upright, once or twice

Defrosting an upright freezer

a year for a chest, and the whole process should take less than an hour.

The most important point to remember is that it is completely unnecessary to run stocks down. If that were the case, it would be a constant process. Certainly defrost before a monthly re-stocking, but never deliberately run stocks down, as that would defeat the whole object of having a freezer.

1. Choose a reasonably cool day or an evening. Disconnect from mains.

2. Remove the food and stack in carry-cold sacks and cover with a blanket. If convenient, use the refrigerator for ice-cream and cream cakes.
3. Remove baskets, shelves or partitions where possible.
4. Place one or two bowls of hot water in the bottom of the freezer and allow to steam for 10 minutes or so.
5. When the ice is soft, gently remove with a plastic scraper or dustpan. Never use metal instruments, as these could damage the cabinet surface.

A thermometer is a useful freezer accessory

6. Do not let the ice run to water, as a bailing-out process is laborious. A little mess is inevitable with an upright freezer, but plenty of newspaper surrounding the cabinet will help.
7. Wash out the interior with a mild solution of bicarbonate of soda, rinse and dry well, particularly in the corners.
8. Reconnect the freezer, replace washed accessories and re-stock. This is a good opportunity to rotate stock and make out a shopping list for your next trip.

A neatly stocked, clean freezer will compensate for an hour's hard work and cold hands. Inevitably, one also finds a few lost packages which turn the chore into a mini-treasure-hunt.

Power Cuts

We might as well accept the fact that power cuts are now a part of everyday life—fortunately, they don't occur too frequently, but for innumerable reasons they will always be with us.

Understandably, they concern the new freezer owners, but the fact is that, so long as certain precautions are taken, the food is perfectly safe in the freezer for 24 hours.

Rarely are we subjected to power cuts of that duration, but should you sustain a longer breakdown all is not lost. Much of the food can be refrozen in one form or another.

Happily, power cuts occur generally in the winter, and so the outside temperature is favourable. Also, many heating appliances are dependent on electricity in some way, and so the house is cool. Power cuts are unpleasant and inconvenient, but the freezer is the least of your worries.

Prior warning is usually given if the power is to be off for more than about an hour. So here's what to do in advance:

1. Move the freezer away from any heat which may be provided by non-electrical means.
2. Switch on to fast-freeze for a couple of hours.
3. Move ice-cream and cream cakes to the bottom or back of the cabinet, stacking the more dense items such as meat on the top or at the front.
4. Ensure that the freezer is well stacked and fill in the gaps with boxes and newspaper to cut down the air circulation.
5. Cover the freezer with blankets—but keep one or two for yourself.
6. On no account open the lid or door until the returning power has allowed the freezer to resume a normal temperature.

If unfortunately you are involved in an extended power cut, chances are that some of the food—if not all of it—will thaw after about three days. Much, however, can be salvaged.

1. Vegetables can be refrozen only if cooked thoroughly and used in dishes. There are numerous possibilities here: vegetable stews and soups are relatively easy and many different varieties could be tried. Also, vegetables can be cooked, puréed and refrozen. Vegetable purée is a useful addition to many dishes for extra flavour and thickening.
2. Raw meats and fish can be refrozen if cooked first. The roasting joints should be cooked, and, when cold, sliced and frozen down for use either in salads or in sandwiches, or frozen in meal portions with gravy. Just reheat to serve. The cheaper cuts, stewing and braising, can be made into prepared meals, together with some of the thawed vegetables. Many varieties of stews and casseroles can be made and refrozen.

   Cooked, refrozen fish does not sound too appetising, but it can be quite versatile. Cod, haddock, plaice and other white fish can be made into various fish flans, fishcakes and casseroles.

   Smoked fish can be utilised to make some fish starters, such as kipper or haddock pâté. Shellfish cooked in an interesting sauce always makes a tasty meal.

   With many of us, fish recipes do not spring easily to mind—so dig out a good recipe book and browse through.
3. All fruits can be refrozen, but this should be done carefully. On thawing, they will have lost some of their juices, and if refrozen in the same pack they will form a solid mass. Remove them from the original pack and freeze free flow—i.e. freezing individually and repacking when frozen. Alternatively, the fruit could be stewed or made into pies, but this is not necessary. As you will be coping with all the other items which *do* need cooking, do not worry too much about the fruit.
4. Bread, cakes and sweet pastry dishes can all be refrozen. Take care to freeze them flat so that they won't damage.
5. Prepared meals, cooked meat or fish must not be refrozen, but may be reheated and consumed. I would suggest that you throw a party to celebrate the return of the power supply and give all the neighbours a jolly good feast.
6. Ice-cream and products containing fresh or synthetic cream must be discarded and never refrozen. Don't take

any chances with these. Throw them away immediately.

So perhaps a long power cut is not as disastrous as it may at first appear. Some foods can be refrozen, others cooked and frozen down as a different food. Remember that when food is cooked it changes its flavour and texture completely, and thus is a different food and perfectly safe to refreeze.

Some food will have to be discarded, but the cost of those should be only a small percentage of the contents of an average family freezer.

Part Two
Filling the Freezer

1. FROZEN FOOD

Without a doubt freezing is the best method of food preservation. The food is preserved in its natural state with only minimal changes in flavour, texture or nutritive value. Bottling, canning and drying serve a very useful purpose, but the characteristics of the food alter during preparation and storage, producing a food quite different from the original.

Freezing is a method of holding food in a state of suspended animation. Once food is introduced into a low temperature, tiny ice crystals form within its structure and almost all bacterial growth and enzyme activity are arrested until the food is required and removed from the freezer. No part of the food is destroyed and thus on thawing it is treated as a fresh product. The secret of good freezing is the speed at which it is done—the faster the process the smaller the crystals, preventing any major destruction of cell tissues. However, a certain amount of tissue breakdown is inevitable, particularly in foods which have a high water content: this is recognisable on thawing, the food going a little limp and the liquid content "dripping". Slight texture changes will thus take place, but in most foods it is unrecognisable and has no effect on flavours or nutritive value.

Although freezing has little effect on the flavour of food, it would be ridiculous to claim that the flavour of frozen food is always first-class. The freezer is a great asset to any home, but it can't work miracles. The flavour of the food on eating will be almost identical to that before freezing. Certain enzyme activity will continue, but, so long as storage times are adhered to, flavour changes are undetectable. The secret lies in storing only best-quality produce whether home frozen or from the supermarket. The same product will come out of the freezer as went in, and so if the flavour is inferior blame the produce, not the freezer.

Freezing has little effect on the nutritive value of food,

A wide variety of pack sizes is available for your freezer

except perhaps causing a slight loss of one or two vitamins, but these are lost to a much greater extent in other methods of preservation and so here freezing is by far superior.

Having established that freezing is the best method of preservation, how does frozen food compare with its fresh counterpart?

The first important thing to do is define what is meant by "fresh"—it seems to have acquired many variations in meaning over the years. So far as fruit and vegetables are concerned, after one day they are no longer fresh. Fish certainly should be eaten within a day of landing and meat within a couple of days of arriving at the butcher's shop.

Having said that, there is nothing to compare with fresh food. It is superior in all respects, but unfortunately there are disadvantages. Much of the produce is seasonal and for most of the year unobtainable or extremely expensive. Transport difficulties hinder regular "fresh" deliveries from the coast, abattoir or farm, and the pure economics of running a business prevents the shopkeeper from maintaining this superior set of standards.

So, for the majority, only second-grade "fresh food" is available. Fruit and vegetables are in the main two or three days old before they reach the shops. Acquiring fresh fish today is rather difficult. Most of it is frozen at sea or on landing and transported in a frozen state to the fishmonger's slab. It is then thawed and sold as "fresh". Meat could be with the butcher for days before purchase—certainly it is kept under hygienic conditions, but it's hardly fresh. Freezers are increasingly being used to store meat, particularly offal, and, of course, imported meat by necessity has been frozen.

This conversation is typical of many I've had with customers about frozen meat:

"Have you tried frozen lamb?"

"Oh, no, wouldn't touch it. We only eat fresh."

"Don't you find English lamb expensive?"

"Oh, we don't have English, always New Zealand—the taste is so much better."

As I've said, nothing compares with really fresh food, but the difficulties in obtaining it and the expense would make it impractical for the majority.

2. FREEZING HOME-GROWN PRODUCE

When you have invested in a freezer it is important that it is used sensibly to obtain maximum economy and convenience. If the machine is half full of runner beans or last year's windfall apples then surely it is not being used to its full potential.

A freezer costs 20p+ per week to run, and so although the windfall apples cost only your own preparation time they are probably now costing 5p to 10p per week to store. That isn't economy by anyone's standards and it is severely limiting the variety which is one of the major attributes of the freezer.

I am not suggesting that garden produce should not be frozen —in fact I'd be the first to proclaim its merits and it is dealt with in detail in Part Four—but I am suggesting that it should be planned and carried out sensibly to obtain maximum benefits.

Grow and freeze a wide variety of fruit and vegetables and preserve quantities suited to your family needs. Don't try to store enough to last until next season as storage costs make it uneconomical. A two- or three-month supply of most items is a recommended guideline and only the produce that freezes well should be selected.

There is nothing quite like eating one's own produce—the devotion and hard work lavished on it somehow improve the flavour 100 per cent—but in reality a good-quality commercially frozen product is very comparable and the range far more extensive. Therefore, as always, a compromise is necessary: by all means practise your skills of gardening and freezing but sensible planning will in the end provide the economy and variety.

3. BULK BUYING TO FREEZE

Fruit and vegetables

Not everyone is fortunate enough to be able to grow their own produce, and so the next best thing is to buy in season and freeze. Done sensibly this works well, but impulse buying can be dangerous as there are numerous pitfalls.

For example (current market prices at time of going to press):

Peas	
Frozen peas (small pack—best quality)	15p per lb
Fresh peas	10p per lb
Fresh peas—bulk buying wholesale price (10 lb)	6p per lb
After preparation	
Peas	4 oz
Wastage	12 oz
Edible peas	24p per lb

At first glance the wholesale fresh peas looked like a good buy but in fact this was misleading.

Runner beans	
Frozen beans (small pack—best quality)	15p per lb
Fresh beans	16p per lb
Fresh beans—bulk buying wholesale price (10 lb)	10p per lb
After preparation	
Beans	12 oz
Wastage	4 oz
Edible beans	13.3p per lb

Preparing fresh food for the freezer is not always economical

The beans seem to be a good buy but let's take into account preparation, packaging and freezing and compare the cost with a bulk purchase of frozen beans.

Blanching cost for 10 lb beans	1p
Packaging 10 × 1 lb polythene bags and labels, twist ties	5p
Fast-freezing for 10 hours	2p
Add these figures to the price of 10 lb of edible beans	£1.33
Total cost	£1.41
To buy 10 lb of frozen beans	£1.30

The price difference is negligible. Is it worth doing it at home?

Clearly it pays to think before you act—the figures I have quoted cannot possibly take into account all individual circumstances. You may wish to buy the fresh beans because they are a special variety unobtainable frozen; on the other hand you may wish to deduct 5p from my costing because you already have the packaging material.

But beware—buying in season isn't always economical and is very time-consuming. The preparation of 10 lb of beans would involve two or three hours, and so don't go overboard and buy a hundredweight just because the price is right—not only will it consume your time for the next few days but also it will take up too much valuable freezer space.

Home-freezing limited quantities of fruit and vegetables can be profitable and fun, but it should be supplemented to a large extent by commercially frozen products which in the final analysis are more or less the same price, are superior in quality and offer unlimited variety and pack sizes to suit everyone.

Fish

The opportunities of purchasing large quantities of fresh fish are limited unless you live on the coast. (The alternative is, of course, having a fisherman in the family.) Buy according to your family needs and freeze down immediately in family portions.

Meat

Unless you are an experienced butcher do not buy carcasses and attempt to joint the meat at home. It is the job of an expert and the wastage incurred will eliminate any possible saving.

For most families carcasses or half-carcasses are not a practical investment although the initial price-saving may be tempting. Unless all the cheaper cuts are to be used, some of which may be unfamiliar, the saving diminishes and it would perhaps have been more sensible to buy smaller-unit packs more suited to the family requirements.

If you are buying from the butcher, check that the price per pound does not include the whole- or part-carcass weight. The quoted price per pound should be on the jointed weight. There is 30 per cent wastage on animal carcasses and so, if the price seems excessively low, check what you are paying for. To arrive at the price of the meat which is left, don't make the usual mistake of adding 30 per cent to the original price. You must add 30/70ths or 43 per cent—e.g. a 30p/lb carcass gives at best 43p/lb nett price.

Cost must take into account packaging and freezing time which can add a sizeable amount to the overall outlay.

Shop wisely. A half-carcass each of beef, pork and lamb would almost fill an average-sized freezer, and so buy according to your requirements and don't let a few pence saving tempt you into over-stocking.

The pros and cons of home freezing are numerous and in the end it depends on getting good meat at a reasonable cost. I would venture to add here that commercially frozen meat is usually of a higher standard—the freezing process is far more sophisticated than that of the domestic freezer, thus ensuring a better-quality product. Prices are competitive and a wide range of cuts and pack sizes are at your disposal.

Whether you choose to freeze your own or buy ready-frozen meat, shop around, but in the end let quality be the deciding factor.

4. FREEZING HOME-PREPARED FOODS

This is where home freezing becomes really beneficial. It is both creative and profitable—far more so than freezing down the raw materials.

Buying ready-prepared foods whether sweet or savoury is reasonably expensive, because you are paying for convenience. A few selected items are a welcome addition to any freezer to provide a quick snack or meal, but the housewife can very soon accumulate her own stock of prepared meals to serve at any occasions.

But, as always, careful planning and cost analysis are important. The saving on making your own fish fingers would be minimal and yet on a stew or casserole perhaps it could amount to 50 per cent.

A freezer cliché has evolved over the past few years—"Eat one, freeze two"—and it is worth remembering. Make treble quantities and freeze the remainder in two-unit packs. This can be done with casseroles, pies, sweet and savoury flans, soups, pâtés, mousse, cakes, to name but a few.

Many advocate setting aside a day for freezer cookery—it could be a good idea but to my mind it is unnecessary. I prefer to cook up the individual items when I feel so inclined and incorporate the preparation around a meal. Having a freezer well stocked with these items is time- and labour-saving. Little extra time is required in making three pâtés as against one, and yet the time and labour saved for the future are considerable.

Many of the dishes can be made from the stocks in the freezer. Buying frozen food in quantity has initially saved money; transferring it into your own prepared meals is extending the saving. All frozen foods can be refrozen: some require cooking first, others don't. Be creative with your freezer—experiment with new foods and recipes. Cook in quantity to save your own time and yet have a stock of home-made dishes on hand.

Sensible planning is the crux of home freezing—it can be enjoyable and profitable, but on the other hand it can be expensive and disappointing.

In short, high-quality raw materials are widely available at competitive prices; on the other hand prepared foods vary in quality and are generally more expensive. So your time is much more profitably spent preparing the latter rather than the former.

5. COMMERCIALLY FROZEN FOOD

When you have invested in a freezer the last thing you should do is become addicted to frozen food. Certainly the family eating habits will change—for the better, one hopes—and shopping trips become less frequent, but don't be reliant on the freezer to provide every meal. A freezer can be of great benefit if used sensibly, but to let it dominate your way of life would be wrong.

A well-stocked freezer supplemented by your own prepared dishes should be used together with other foods. Bargains are always to be found in the High Street—don't pass them by. Many foods do not freeze well, others are better preserved in tins or packs, and some are better just eaten fresh.

Frozen food certainly has many advantages but it doesn't claim to be a substitute for everything else. Convenience foods play a major part in our eating lives, but don't let them take over.

Shopping for Frozen Food

Three main sources are available:

Bulk delivery.
Supermarkets.
Freezer Food Centres.

All have their advantages and disadvantages.

Bulk Delivery

A delivery service is ideal for those who are housebound or have transport problems. The food is ordered from a price list over the 'phone and delivered within a few days. But the snags must be considered. Ordering from a price list can be misleading

Shopping at a Freezer Food Centre

because you are reliant on someone else choosing your food and so far as meat is concerned this is critical; nor can out-of-stock items be substituted. Awaiting delivery can be a problem as unlike milk the food cannot be left on the doorstep.

When making enquiries together with comparing prices, check if delivery is chargeable and also if deductions are made for large orders. This could be advantageous if you go into partnership with your next-door neighbour.

Reputable companies operate efficient schemes, but the disadvantages appear to dominate.

Supermarkets

Many chains are now devoting cabinet space to freezer packs. The prices in the main are attractive but the choice very limited. One or two items will possibly be a good buy, but for monthly freezer shopping supermarkets are not recommended.

Freezer Food Centres

A phenomenon of the late 1960s, the Freezer Food Centre, is a supermarket specialising in frozen food. The Centres operate on the same lines as a supermarket and that is the main advantage. All the food is on display, a wide variety is offered, and substitutions can easily be made for out-of-stock lines. Choice is simplified by varying pack sizes which help in planning the family requirements for weeks in advance.

To be able to see the food and shop at leisure is an obvious advantage. The main snag, however, is the temptation of impulse buying. One of the aims of the Centre is obviously to create this and one wouldn't be human not to give in to temptation occasionally, but exercise your willpower and try not to overspend on luxury items. Transport could for some be difficult, but as the visits are only monthly or bi-monthly the cost of a taxi shouldn't be ruled out.

The development of the freezer market has escalated during the last few years and the sales of frozen food have grown with it. As the freezer becomes an accepted piece of household equipment so more Freezer Food Centres will spring up. For you, the customer, this can only be a good thing, as competition will help maintain high quality while prices are kept steady. Many of the supermarket chains have plans to enter this field on a large scale and so the outlook appears promising.

I have avoided mentioning the door-to-door sales operations whose schemes involve buying a package deal of freezer and food over two or three years at a high interest rate. My opinion is that these schemes will not be with us for much longer: the public has become wise to the "hard sell" and prefers to shop around. If, however, you choose to consider buying by this method, check how the food delivery service operates and if it is suitable for your requirements.

Planning the Shopping

The first shopping trip will probably be a little bewildering: planning for weeks ahead rather than days is not necessarily easy. It is difficult to discuss this in general terms as everyone's choice and needs vary so much. Where to shop? How often to shop? What to buy? The answers to these questions differ according to family requirements.

Plan out how much you intend to spend according to your normal food bill. Carefully work through a price list and tick off the items required. It will be virtually impossible to stick to it rigidly, but a guideline always helps. Ensure the price list is up to date and once in the store watch out for special offers which could be money-saving.

Compare prices carefully and stock the freezer with the items which are going to save you the most money.

I trust the following example will serve as a guide around which individual needs can be adapted.

Stocking-up for the freezer

Household expenditure breakdown (family of four)

Expenditure £11.50 weekly		
Meat and fish	£3.50	Butcher, fishmonger
Vegetables and fruit	£1.00	Greengrocer
Convenience foods	£1.00	
Dry goods, eggs and bread	£3.50	Supermarket
Household sundries	£1.00	
Milk bill	£1.50	Milkman
Transferring to a four-weekly shopping expenditure £46		
Meat and fish	£14.00	
Vegetables and fruit	£4.00	Freezer Food Centre
Convenience foods	£4.00	
Dry goods, eggs and bread	£14.00	Supermarket
Household sundries	£4.00	
Milk bill	£6.00	Milkman

Expenditure need only be adapted to suit the new routine. If possible, transfer from a weekly budget to a monthly budget—freezer shopping is best done monthly or bi-monthly. Also try to organise the supermarket shopping on the same basis. Obviously many items have to be bought fresh, but the bulk of the shopping you should aim to do monthly.

From the above simplified expenditure breakdown a total of £22 is allocated for monthly freezer shopping. It is important that this money is spent proportionately on the items required according to the family weekly expenditure.

The following example suggests purchases for the first shopping trip.

Suggested Shopping List

Meat & Fish		
	2 Beef joints 3 lb Braising 4 lb Lamb legs 3 lb Pork chops 2 Chickens 3 lb Lamb's liver 2 lb Bacon 3 lb Breaded cod 3 lb Kipper fillets 36 Fish fingers	**£14**
Fruit & Vegetables		
	5 lb Peas 2 lb Beans 5 lb Chips 2 lb Stewpack 2 lb Mushrooms 2 lb Raspberries 1 lb Apricots 2 lb Fruit salad	**£4**
Convenience Foods	1 gallon Ice-cream 5 lb Pastry 1 packet Pizzas—4 1 box Cottage pies—8 1 box Hamburgers—24	**£4**

On the first shopping trip £22 doesn't go very far and this will obviously limit the variety in the freezer. Some of the packs will serve as a two-month supply and so future shopping trips will add to the variety quite quickly. The example shown hasn't in fact saved any money, but has purchased better value for money. The normal four-weekly outlay of £22 has merely been redirected.

To my mind, better value for money is in fact money saved, but I would agree that those who bought the freezer to save money in cash terms have nothing to show. For the first few months a cash saving is not always possible: this can be achieved only with a well-stocked freezer being used properly.

After the initial shopping trips, however, the monthly outlay could be dropped by £1 or so. This will not be easy to estimate as the cost of living fluctuates consistently, usually upwards. Doubtless savings can be substantial, but to show cash-in-hand proof would not be so easy.

If it is within your means to lay out more money on stocking the freezer initially, many advantages can be gained. Instantly you are using the freezer to its best advantage: running costs are kept to the minimum, and from square one you are saving money.

Do not, however, be tempted to buy large economy packs before the item has been sampled. The 20 lb box of peas may save 30p or more, but if the family doesn't like them then it was hardly a bargain. Think carefully before buying that half-pig pack: it's only money-saving if you use all the cuts. For the first few months concentrate on small packs until the family favourites have been established—then invest in the money-saving economy packs.

Ideally £40 or £50 should be spent on the initial shopping trip divided as the suggested ratio, or as required. As with anything new there will be an element of trial and error, but don't be afraid to seek advice from the staff. An honest opinion could be a valuable asset.

Freezer shopping will seem strange at first, but one soon gets used to it.

6. STOCKING A FREEZER

Upright

Packing an upright is, of course, dependent on the model, but the following general guide can be adapted to your own particular freezer.

In an upright you should be able to pack 12 to 15 lb per cubic foot, and so in a 12 cu ft upright it is possible to pack 180 lb of food.

Ideally, reserve a shelf for each of the following: meat, fish, fruit and vegetables, ice-cream and cakes, convenience foods.

A well-stocked upright freezer

Decide according to the family's needs.

It may be necessary to repack some bulk items. A 20 lb box of vegetables may have to be split into 5 lb or even 1 lb units. Smaller items are far easier to pack into the freezer and are also much more convenient to use in the months ahead. When repacking, don't forget to relabel and date. In fact it is a good idea to date all the purchases when you bring them home so that a good stock rotation can be put into operation.

All the bulk meat packs should be double wrapped, and so if an economy pack won't fit on a shelf merely take out all the joints, slip each of them into a separate polythene bag and pack into a shelf. Once you have completed repacking, stack each shelf as carefully as possible to avoid wasting space. Put items not for immediate use to the back of the shelf, making sure that boxes are stacked on top of one another.

Use sliding baskets to their best advantage by having everyday foods stored in them. An upright freezer with two doors can help quite a lot with stock rotation. Pack the top half of the machine with all the foods for immediate use, and in the bottom store the bulk items. Transfer food up to the top weekly and put monthly bulk shopping into the bottom. By adopting this habit you will avoid storing foods longer than the recommended storage time.

If shelf space in the door is available, use for storing small packs such as fishcakes, pâté, butter, etc. Avoïd putting ice-cream, cakes or cream in the door space as this will cut down their storage life. Whenever the door is opened, food in the door space is exposed to the outside temperature, and so care must be taken not to keep it there for too long—one month should be the maximum.

In most upright machines there is no specific fast-freeze compartment, and so any food you are freezing yourself should be placed on the top or bottom shelf, or stacked against the walls of the unit. The coldest part of the freezer will vary according to the make, but the position in the freezer is relatively unimportant. What does matter is that the food is in contact with the freezing coils.

Chest

Bulk items can easily be stored in the bottom of a chest freezer as there is no restriction of shelves. 20 lb per cu ft is the average holding capacity, and thus a 12 cu ft freezer will hold 240 lb approximately.

It may be convenient to repack larger items for future use. Trying to remove ¼ lb of beans from a 20 lb box in the depths of the freezer can become a daily chore.

Use dividers or plastic carrier bags to section up the bottom of the freezer. Also use the fast-freeze compartment as storage space: it is particularly useful for protecting the more delicate items such as cakes and ice-cream.

Use the baskets on the top for small items and everyday needs. Ensure that the baskets slide easily and give adequate access to the bottom without lifting. Most baskets are large enough to hold a week's supply of food. Advance weekly planning could help keep the freezer organised and eliminate constant dives into the depths for odds and ends.

A chest freezer can become chaotic if not used properly, and

A selection of useful freezer aids available from most department stores and Freezer Food Centres

so a little forethought and planning are required. Stock books, record cards, etc., are claimed to work miracles for the disorganised. Frankly I think they are a worthless investment. Mine ended up more disorganised than the freezer. I'm sure they are excellent for efficient people—but would they need them anyway? My memory has served me very well up to now, but I'm careful to label and date as a precaution.

7. STORAGE OF FROZEN FOODS

Possibly the most confusing aspect of freezing is the storage. How do I know how long the food has been frozen before I buy it? How long can foods be kept in my freezer? What happens if I overrun the storage life?

First of all the temperature in the freezer must be correct. The suggested storage temperature is –18°C (0°F), but it is advisable to hold that temperature a little lower, i.e. –20°C (–4°F), to allow for temperature fluctuation when opening the door or lid. If a thermometer isn't supplied with the freezer it is an investment to buy one. It will help you to check at a glance each day if the freezer is running normally and indicates when the thermostat setting needs adjustment.

Once the thermostat is set correctly according to the location of the freezer it should need adjusting only occasionally, perhaps in a rare heatwave or a very cold spell. The thermostat on a freezer works exactly like that on a refrigerator. Some work on a number system, the higher the number setting the lower the temperature; others are marked "minimum", "normal" and "maximum", which is self-explanatory. The instruction booklet with the freezer will give the necessary directions.

Having established that your freezer is running at the correct temperature, we can now go on to discuss the questions posed at the beginning. From the outset it must be made absolutely clear that no harm can come from eating frozen foods so long as they are held at the correct storage temperature, even if the storage life has been exceeded two or even three times. There are various types of bacteria, enzymes and micro-organisms in food, their activity being strictly related to temperature.

1. Food Poisoning Bacteria

These are very active at room temperature, but in refrigerator temperature grow only slowly. Below refrigerator temperature they are dormant and so cannot possibly affect the food in any way.

2. Micro-organisms

These have little harmful effect on food—in fact they are used to cultivate and mature many foods, for example cheese and yoghurt. They show themselves in the form of yeasts and moulds.

Micro-organisms are active at temperatures down to –7°C (20°F). Below that they are dormant, and so they cannot have any effect on foods in the freezer.

3. Enzymes

Enzymes affect the taste and texture of food, but have no harmful results at all.

Their activity does continue at –18°C (0°F) and below but it is negligible. So enzymes will affect the taste and flavour of frozen food over a period of time but in no way render the food inedible. Only the most sensitive palate would detect a great change.

When buying frozen food it is obviously impossible to tell how long the food has been frozen.

Commercially frozen foods are stored at very low temperatures, where enzymes are inactive. So in fact foods could be stored at these excessively low temperatures for years without any change in the food whatsoever.

As you might imagine, in the reputable food chains checks are kept on foods during transportation, and on arrival at the retail outlet they are often stored in large cold stores maintaining a low temperature. Their life actually in the cabinets from where they are purchased is at the most a matter of weeks and here again a check on stock turnover is carried out to ensure that the food you purchase is in prime condition.

Most major companies, both suppliers and retailers, have well-established quality control departments, whose function is to ensure that the produce bought by the customer is of good quality and in perfect condition.

Seasonal produce is, of course, frozen down during specific weeks in the year, so that many of the vegetables you buy in February or March were frozen during the previous year, but because of the strict control of temperature those vegetables taste exactly the same as the ones eaten in the previous August or September. Many items are shipped in from abroad, but the same rules apply.

So after you have purchased some frozen food how long can it be stored? Unfortunately, there are many schools of thought here, which makes life rather confusing for the housewife. Many packs do state a storage life, but on the whole they are rather restricting, as most state between one and three months. The supplier obviously wants you to eat his produce in prime condition and so he ensures this by stating a very conservative storage time. In many cases you wouldn't want to keep the food for a longer period anyway. A 2 lb pack of peas wouldn't last the average family a week, and so to state that its storage life is three months is more than ample, but on other produce this may not be so although it should not deter you from buying.

A pack of chicken pies may state a storage life of one month, and yet for your family to get through 24 pies in a month would mean eating them almost every day. So do bear in mind that quoted storage lives can be extended. The flavour of the food may change slightly but it would be undetectable to most.

If you do keep foods for well over the storage time (i.e. sausages for a year, perhaps because they were lost only to be

Packaging materials and containers can be purpose-bought, but good use can be made of margarine and yoghurt tubs, etc.

revealed on defrosting), then it is advisable to use the produce incorporated into a dish rather than served on its own. The flavour change, if any, would then go undetected. So long as the foods are stored at the correct temperature and are not subjected to continual temperature fluctuations, most will store for a reasonable period of time. Losses of flavour, colour and texture will result from excessive storage periods, but they are unnoticeable in most cases.

It is wise to turn over the contents of the freezer two or three times a year if it is to be used to its best advantage, both for convenience and economy. But with certain items this may not be practical, and so try to adhere to the suggested recommended storage times listed on pages 42–43.

Once removed from the freezer the foods should be used as soon as possible. Store and use as fresh, but bear in mind that frozen foods do tend to deteriorate a little faster than their fresh equivalents. Many foods, of course, can be cooked and used straight from the freezer, which is obviously the best method where applicable.

Recommended Storage Times from Time of Purchase

Meat and Poultry	Months
Beef and lamb	10–12
Pork and veal	4–6
Offal	3–4
Sliced bacon and other cured meats	2–3
Ham and bacon joints	3–4
Chicken and turkey	10–12
Duck and goose	4–6
Venison	10–12
Rabbit and hare	4–6
Sausages and sausage meat	2–3
Minced beef	3–4
Fish	
White fish	6–8
Oily fish	3–4
Fish portions	3–4
Shellfish	2–3
Fruit and Vegetables	
Fruit either with or without sugar	8–10
Fruit juices	4–6
Most vegetables	10–12
Mushrooms and tomatoes	6–8
Vegetable purées	6–8
Dairy Produce	
Milk (homogenised only)	3–4
Cream	6–8
Yoghurt	3–4
Butter—unsalted	6–8
Butter—salted	3–4
Eggs	8–10
Cheese—hard	4–6
Cheese—soft	3–4
Ice-cream and similar products	3–4
Prepared Food	
Ready-prepared meals—highly seasoned	2–3
Ready-prepared meals—average seasoning	4–6
Boil-in-the-bag meals	4–6

Cakes	4–6
Bread—all kinds	2–3
Sandwiches	2–3
Bread dough	2–3
Other yeast products and pastries	3–4

This list relates both to foods purchased ready frozen and to those purchased fresh for storage in the freezer. Part Four is devoted to the preparation and storage of home-made items.

Part Three
Buying and Cooking Hints

1. MEAT

Introduction

Sunday lunch, 1675—"one roast ox, one roast sheep, six pheasant . . . ." A little different to the roast and two veg of Sunday lunch today! The cost of living has restricted our eating habits over the centuries and today the average family is content to eat its way through an animal over a period of months.

We have progressed from the home-reared spit roast to the butcher's shop. Various preservation methods have been accepted over the years—smoking, tinning, tenderising—and now frozen meat is rapidly cornering a major part of the meat market. But is frozen meat really so revolutionary? No, of course not—last century imported frozen meat was on sale in Britain, but in the main the housewife was unaware of it. It is only during the past few years that frozen meat has been widely available and for some the problems of selection have deterred buying. Frozen meat looks so much different from that in the butcher's window. To unaccustomed eyes it appears unappetising—mainly because the freezing process darkens the colour. It is hard to change our shopping habits overnight—the problem is one of unfamiliarity. The eating quality, which is, after all, the important part, is comparable to that of meat bought fresh. The only difference is its storage convenience for the freezer owner—and, of course, price!

Buying Frozen Meat

Any item which accounts for 30 per cent of the household expenditure on food must be worthy of considerable thought before purchase. It would, however, be wrong to buy meat purely on price; inferior meat even at exceptionally low cost is rarely good value for money—it is wiser to pay an extra pence or two per pound for good quality whether it be stewing meat or fillet steak, particularly where large quantities are concerned.

Buy only from a reputable supplier and try smaller quantities before buying in bulk. Store a wide selection of cuts according to your family requirements but sufficient only for a two- or three-month supply. Although meat may appreciate in value, freezer space is a much greater asset. When you have established your requirements through usage over a couple of months, re-stocking may involve the purchase of one or more bulk economy packs. The contents of these vary—that of a half-carcass of pork or lamb is obvious, but beef packs differ enormously and so before buying check the contents. There is no doubt that these packs are money saving but only if all the cuts are put to good use—if some will be wasted then don't buy. For the more adventurous with a family prepared to try new dishes, bulk packs can be fun as well as money saving.

Smaller meat packs will be the best buys for many. Joints are packed in twos or threes. The steaks, chops and cheaper cuts are in 3–5 lb units—all the meat should be packed free flow for easy use. Mince and steak and kidney should be packed in 1 lb or $\frac{1}{2}$ lb units. Select carefully to ensure that the quality is maintained through the pack. Although the price saving is not so attractive for the small family these packs are in the long run a better buy.

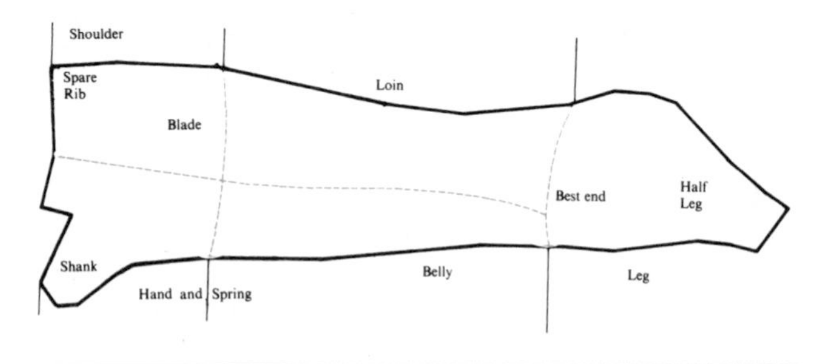

Pork—average weight of a half-pig pack 40 lb

Leg	A succulent roasting joint usually cut into two or more joints for normal family requirements.
Belly	Usually sold as belly slices—grill, fry or roast. Alternatively, salt and boil in one piece.
Hand	Best served as a roasting joint boned and rolled but can be boiled or pickled.

Spare rib	Joint or chops—roast or grill accordingly. Excellent braised for oriental dishes.
Blade bone	Economical roasting joint on the bone. Bone and stuff if preferred—for extra flavour.
Loin	Can be cooked whole as a roasting joint, but more usually sold as pork chops—fry, grill or braise.
Half-head	Always included in a half-pig pack—used for brawn or pâté.

The contents of a half-pig pack

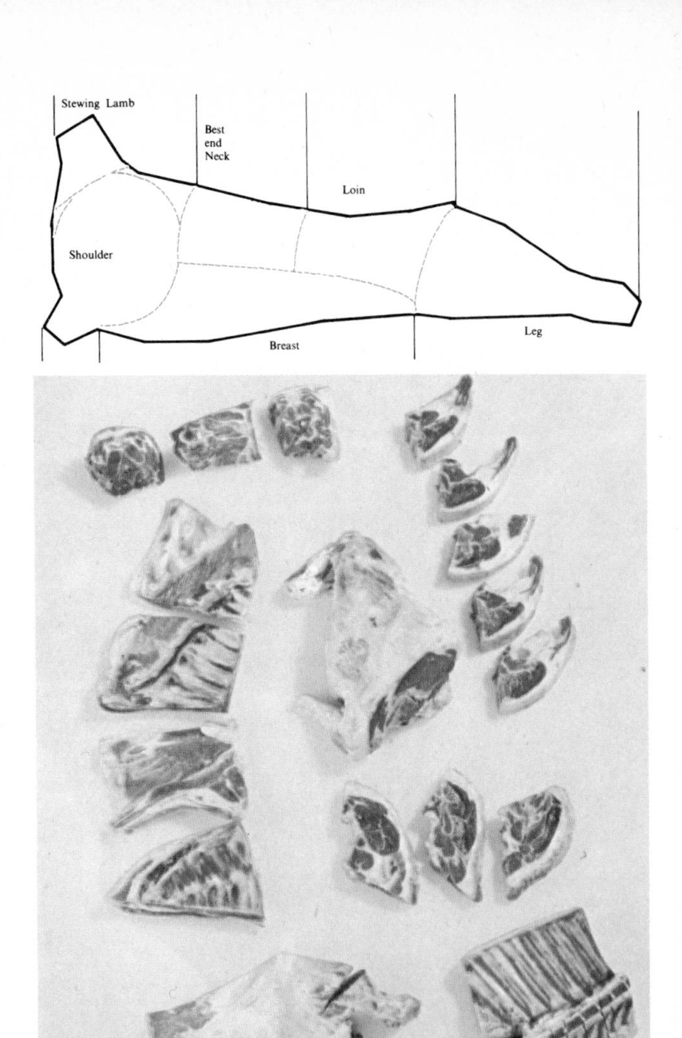

The contents of a half-lamb pack

Lamb—average weight of a half-lamb pack 16 lb

Leg	Roasting joint on the bone.
Breast	Economical roasting joint. Bone and stuff for added flavour.
Shoulder	Roasting joint—can be used for special dishes or casseroles.
Scrag end and middle neck	Economical stewing cuts—long cooking time required for best results.
Best end neck	Reliable roasting joint on the bone. Can be cut or used as braising chops.
Loin	Can be roasted as a joint but generally used for loin and chump chops—best grilled or fried.

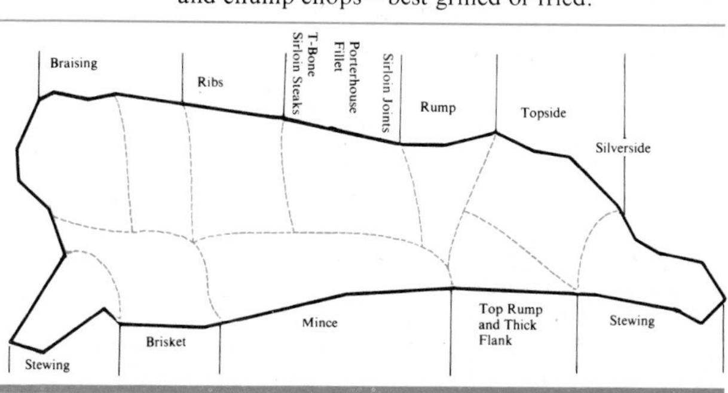

Beef—average weight of an economy beef pack 30 lb

Rump	Steak for grilling or frying.
Topside	Roasting joint—lean with added fat for cooking. Tender and juicy.
Silverside	Can be roasted but better boiled or pot roasted. Flavoursome but can be dry.
Shin	Needs several hours' stewing for best results. Use for stews, casseroles, pies.
Top rump	Roast or braise. Good flavour.

Flank	Usually sold as minced beef for use in economical meat dishes.
Brisket	Slow or pot roast. Excellent flavour and tender if cooked slowly.

Cooking Frozen Meat

Roasting Joints

To get the best results, joints of beef, pork and lamb should be cooked straight from the freezer.

The contents of an economy beef pack

There are two main reasons for this. Firstly, it is more convenient and, after all, that is what the freezer is for. It is no longer necessary to think 24 hours ahead. Secondly, and of much more significance, the results are better. When thawing a joint, meat juices are always lost; these valuable nutrients are often thrown down the sink and the joint is less moist as a result. By cooking from frozen, the juices are retained in the meat.

The use of a meat thermometer is strongly recommended. Although the suggested cooking times have been thoroughly tested, individual tastes differ. Some prefer beef rare, others well done—meat is cooked to perfection only if it suits one's taste. A meat thermometer helps achieve this.

Method

1. Pre-heat the oven to 180°C (350°F), Gas Mark 4.
2. Place the joint in a cook-bag and seal loosely with a twist tie.
3. Put the bag into a shallow roasting tin and position centrally in the oven.
4. Calculate the cooking time:
   Beef 50 minutes per lb
   Lamb 60 minutes per lb
   Pork 60 minutes per lb
   e.g. 2½ lb beef joint 2 hours 5 minutes
   3¼ lb pork joint 3 hours 15 minutes
5. Approximately 15 minutes before the estimated end of cooking time plunge the meat thermometer through the cook-bag into the middle of the meat and take a reading.
6. Calculate the remaining cooking time accordingly. Repeat the process as often as necessary to achieve a perfectly cooked joint.
7. If a crisp exterior is required, i.e. pork crackling, slit the cook-bag 20 to 30 minutes before the end of cooking time.
8. When cooked, slit the bag and remove the joint. Carve and serve. The liquid remaining in the bag is pure meat juice and makes excellent gravy.

A joint of meat being placed in a cook-bag

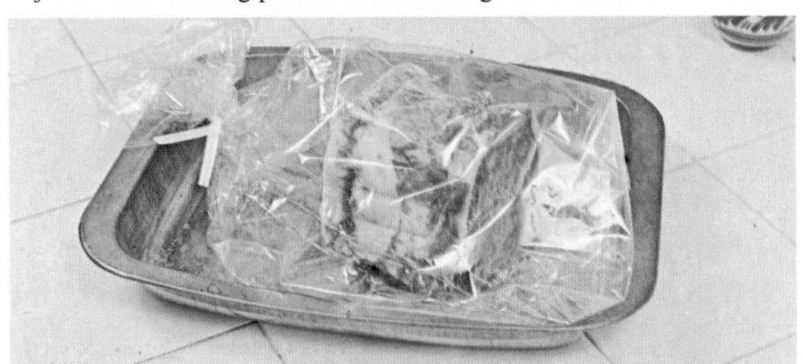

Sealed and ready for the oven

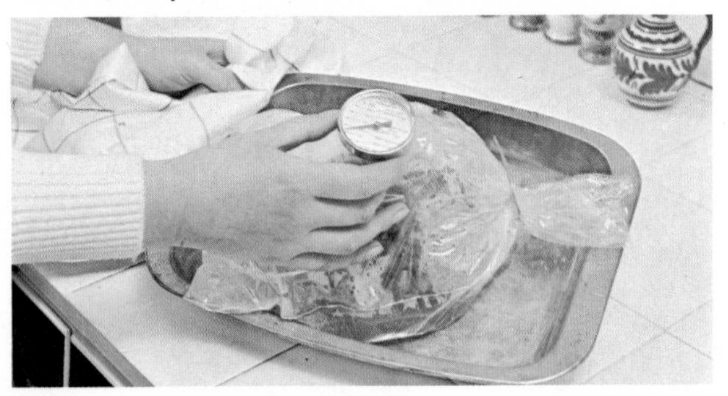

Testing the joint with a meat thermometer

The use of a cook-bag isn't essential but has many advantages. The two major ones are that the meat is self-basting and thus cooks in its own juices, and also as the meat is totally enclosed the oven is kept clean.

If a cook-bag is not used the meat must be well covered and basted frequently to prevent the surface drying. Also the cooking time may need extending by approximately 5 minutes per pound.

Cheaper joints such as brisket or breast of lamb are certainly improved by cooking from frozen—a reduced cooking temperature and an extended cooking time are recommended. For example:

Brisket	150°C (300°F)	60 minutes per lb
Breast lamb	150°C (300°F)	70 minutes per lb

Occasionally, however, larger joints require boning and/or stuffing and rolling—obviously thawing must precede this process. Cook in the traditional way.

Grilling a pork chop, tomatoes and mushrooms from frozen

Steaks and Chops

The same principle applies. Cooking from frozen is more convenient and gives a more succulent result. As frying is a more moist method of cooking, it is perhaps preferable to grilling, but both methods can be recommended.

Seal the steak or chops under a hot grill or over a high light until browned slightly on each side. Reduce the heat and continue cooking. Baste frequently to keep the surface moist. It is difficult to estimate accurate cooking times as the thickness of the cut and the individual palate vary so much. A general guide would be just less than twice the cooking time for fresh cuts. For example:

Fresh steak	5 minutes
Frozen steak	8–11 minutes

Frequent basting is essential but covering the meat is not recommended as the pressure extracts the meat juices.

Inexpensive Cuts
Stewing, braising and mince are in the main used for stews, casseroles and pies, and can in most cases be used from the freezer. If the meat is to be fried before being mixed with other ingredients, place in hot fat and fry gently until thawed and well browned. If cutting or dicing is necessary it can be done at this stage. Alternatively place the meat into hot stock and simmer gently until separated. Partial thawing will be necessary if the meat is to be coated before cooking. After these initial stages cooking should continue as normal.

Cooking meat from frozen has many advantages but it may not suit everyone. Thus a word or two about thawing meat. It must always be done in the refrigerator—unwrap the meat, stand it on a plate and cover with a clean wrapping. Allow 24 hours for all cuts and small joints—joints over 5 lb may need 36 to 48 hours to ensure complete thawing. Use the drip both to baste the meat and for an accompanying sauce or gravy.

2. POULTRY AND GAME

Introduction
The range of meats under this heading is vast. To most of us many are unfamiliar as they are rarely seen in the High Street. With the growth of the Freezer Food Centre, however, it is probably likely that more of these will be available to the average family, along with the popular chicken and turkey.

Game is protected by law and may be shot only during specified months of the year. The freezer will eliminate the seasonality and imports will probably boost the availability.

So watch out for pheasant, partridge, pigeon, venison and hare in your local Centre—they are not always expensive and can be delicious if used in old traditional recipes. As poultry is by far the most popular choice, the rest of this section is devoted to it.

Buying Poultry

Identification	
Poussin	Baby chickens 4–6 weeks old, delicate in flavour but not a lot of meat. Serving: 1 per person.
Roasting chicken	Year-old cocks and hens, moist tender roasting joints. 3 lb: 3–4 portions.
Hens	Boiling fowl, which requires long, slow cooking. Good flavour and an economical buy for casseroles and pies. 3 lb: 3–4 portions.
Capons	Young cockerels, specially reared to give a plumper bird. They are roasting joints and are good value for money because of the meat content. 5 lb: 4–6 portions.
Turkey	At their best between 7 and 9 months old. The hen bird is usually found to be more tender than the cock. 8 lb: 6–8 portions.
Duck	Rather fatty, but an excellent flavour. A little stronger than other poultry. 4 lb: 2–3 portions.
Goose	Eaten usually when 4–5 months old. As with ducks, rather fatty and stronger in flavour. 8 lb: 4–6 portions.

It is easy to miscalculate portion servings on poultry. When buying, always overestimate rather then underestimate to avoid disappointment especially if entertaining. Left-overs can always be refrozen for future use. As well as whole chickens, chicken portions—drumsticks, wings or chicken quarters—can be obtained. These are usually sold in 3–5 lb units; larger packs are better value for money if storage space permits—these are usually a selection of random cuts in 15–20 lb boxes. For some families portions are more useful, and so if the intention is to serve this way buy chicken portions rather than joint your own. For entertaining large numbers they are simple to prepare and good value for money.

For Christmas and Easter buy the turkey a couple of months in advance as it would be idealistic to assume that price fluctuations do not occur on frozen food. Also, by buying early you can ensure a good selection of weights.

Roast pheasant with apricots and brandy

Cooking Poultry

It is essential that poultry is completely thawed before cooking. The reason for this is that without doing so it would be extremely difficult to estimate if the bird was cooked thoroughly. The flesh could well be cooked to perfection, but the interior carcass still frozen. (Also it is impossible to remove the plastic bag of giblets while frozen!)

Thawing should always be done in the refrigerator if possible. Fast thawing always produces excessive drip, reducing the flavour of the bird.

Chicken and mushroom casserole—an easily prepared dish

Chicken/Capons		
	Under 3 lb	12–24 hours
	Over 3 lb	24–36 hours
Ducks		
	3–5 lb	24–36 hours
Geese/Turkeys		
	Under 8 lb	24–36 hours
	8–14 lb	36–48 hours
	14–20 lb	48–60 hours
	Over 20 lb	72 hours

If it is necessary to thaw at room temperature, these times can be reduced by a third.

Never thaw out in water—the moisture lost could be such that the bird is dry and tasteless. Once thawed, cook in the normal way. Chicken portions can be cooked from frozen as there is no hollow carcass to restrict heat penetration. The portions should be fried or grilled in butter—frequent basting is essential to prevent excessive drying. Test by means of a skewer: a creamy secretion is proof of thorough cooking; a reddy liquid is a sign of undercooking.

3. CONVENIENCE FOODS

Introduction

Convenience foods are a necessary part of the busy housewife's everyday life, and yet you will hear many denounce them as offensive and nutritionally worthless. But I wonder how many of those housewives who disdainfully express these views have made their way through the past decade without sausages, sliced bread and teabags, and I've yet to find home-produced baked beans!

All frozen foods are convenience products—no preparation, no wastage, just cook or thaw as required. The fact that they are ready prepared doesn't mean they are inferior—in most cases quite the reverse. Neither does it necessarily mean expensive—mass production and bulk buying keep costs down. Convenience foods have been with us since the village dairy started selling cheese and butter, and so the packaged meals of today are certainly not revolutionary, merely a progression.

To discount them as "expensive and nutritionally worthless" is an unfair generalisation—of some that may be true, but on the whole it is not. Their preparation is similar to that in a domestic kitchen, only on a much larger scale, and the ingredients are almost identical. But buy only convenience foods of a well-known brand name if unsure. To claim they are expensive is relative to the price you put on your own time. If the housewife has time to make her own fishcakes and pizzas, then that's fine, but for most it is very time-consuming and the extra cost incurred in buying them is as nothing compared to the time saved. On the whole, convenience foods are nutritious and good value for money if bought in quantity.

The acceptance of convenience foods is merely an attitude of mind. The choice should be determined not only by family likes and dislikes but also by what is of most convenience to you.

Most of us would be reluctant to admit that our culinary skills are not as advanced as we would like to think, and sometimes one wants to impress. Why shouldn't you dig out a pâté and an authentic French cuisine masterpiece from the freezer? True, you could have made it for two-thirds of the price, but cost isn't everything and to know that the meal is going to be a success before you even start to prepare it is preferable to panicking about a disaster when it's too late. And who's to know—garnished skilfully you can disguise even a fish finger.

Convenience foods serve two main purposes. They provide

Duck pâté fresh from the freezer

quick and easy snacks and meals for all the family which are particularly useful in your absence. Also they are invaluable for that special occasion when time is perhaps short and success essential.

Stock up with the foods to suit your requirements and supplement with home-made dishes a supply of which is easy to build up over a period of months. If used sensibly they add variety to the family diet and are not over-priced if selected carefully.

Buying Convenience Foods

It is difficult to categorise ready-prepared convenience meals and snacks as the range is vast and increasing daily. However, they seem to split into two main sections, everyday foods and those for special occasions.

Everyday Convenience Foods

Before buying in quantity try smaller packs if you can. However, this may not be possible as many suppliers pack only in quantity. So if sampling is not feasible ask the advice of the shop staff or take recommendations from friends. It is wise at first to buy familiar brand names but do bear in mind that many others, although not established in the domestic market, have for years supplied the catering trade. Buy for quality rather than price and do not try too many different types at one time. Evaluate each product and if it pleases the family and your purse then stick to it. Don't be tempted to try a cheaper brand to save a few pence unless it is recommended by a reliable source, as you may well be disappointed. A little trial and error is inevitable at first until the family favourites have been established.

Fish fingers, a creation of the 1950s, are now part of our staple diet. They are in fact very nutritious and in today's world of rising prices relatively cheap. They are made from whole pieces of fish, unlike many other fish creations such as fishcakes. So when buying fish products check the fish content—they may be cheaper than the fingers but are they as good? Fish fingers and their counterparts are enjoyed by many children who wouldn't eat fish in any other way—so if only in that respect they serve a worthwhile purpose.

Burgers come in various shapes and sizes—some good, some not so good. Don't be misled by the terms "hamburger" and "beefburger"—they are very similar products and only the connoisseur could tell the difference. So don't buy a pack of each, as that would be a waste of freezer space. Burgers are usually served with chips or in a bread roll, but for a change try them in a stew with an interesting sauce. It's surprising how many exciting meals can be created from a simple burger.

Pies and puddings are always a good stand-by, although, while it can be said of most that the pastry is good, the amount of filling is always the questionable factor. The meat content can be disguised by a highly flavoured gravy and thus these items are not particularly good value for money. On these foods perhaps more than any other, buy for quality—cheaper products are certainly inferior.

Sliced meats, stews and meals on a tray are handy to have in the freezer, but my opinion is that they are relatively expensive to buy and very cheap and easy to do at home. My main criticism of the bought product is the scant portion, and as these are generally bought in one- or two-portion sizes they must be a disappointment to a hungry family. Some are better than others, but until standards improve I would recommend that this is where home-freezing can show profits while providing a much more satisfactory meal.

For entertaining, packed lunches and picnic meals, a selection

Duckling à l'orange is available in convenient ready-frozen packs

of sausage rolls, pastries and pies is extremely useful, but don't buy in very large quantities except perhaps in the summer when outdoor eating is enjoyed. Again, quality varies—so choose carefully.

Special Occasion Convenience Foods

"Special occasion" doesn't necessarily mean Christmas or entertaining. Sometimes the family may wish to celebrate or just try something different. For all these occasions there is a wide variety of special prepared meals usually of Continental or Oriental origins. While some are expensive, many surprisingly are not if you consider the time allocated to preparation and the spices and liqueurs used which are out of the budget of many housewives.

French cuisine requires skill and patience to prepare—dishes of high quality are obtainable from most Freezer Food Centres supplied by producers intent on quality, price taking very much a secondary place. These are ideal for entertaining: the portions are generous and the flavour superb. Authentic pâté can be obtained frozen, but ensure that you are buying pâté and not paste or slicing sausage. Advertising can be misleading sometimes.

Indian and Chinese dishes are becoming increasingly popular and frozen meals usually sold in one-portion packs are a useful stand-by. They are not an economical buy, but the portions are usually ample. The Indian curries in particular, I suggest, are a good buy, preferably without the rice. Rice is so easy to prepare while the curry is heating that I begrudge paying the price for it frozen. The range of Chinese meals is rather disappointing although some are extremely good. Before buying, evaluate the cost: is it cheaper to have a Chinese meal out or collect to eat at home instead of buying a boil-in-the-bag? The answer may be surprising.

Pizza and other Italian dishes are a good buy. Pizza is not a particularly practical snack to make at home unless done in large quantities, and it is even more impractical if one considers the wide variety of frozen ones available. Quality and price are variable, but in this instance I would suggest buying for price rather than quality. The dough base and sauce are usually good. The difference is in the amount of topping, which can easily be augmented when cooking. Grated cheese, tomato, herbs, sardines and a whole variety of extra toppings can be added—an excellent way of using left-overs while satisfying individual tastes.

Pasta dishes such as spaghetti, lasagne and ravioli are finding their way on to the market. Frozen varieties should be compared with tinned before buying, but the former, although a pence or two more, are a better product on the whole.

It would be impossible to mention the whole range of convenience products and anyway such a list would soon be out of date.

The most important point is to buy items which are going to be of convenience to you. If it is as easy to make them at home for less cost then do so, but for all of us some prepared foods are essential. So buy and select those which suit the family, and don't forget to have a few with easy-to-follow cooking instructions so that you can have a day off once in a while and let someone else do the cooking.

A bought frozen pizza base with a filling home made from left-overs

4. VEGETABLES

Introduction

Vegetables contain important vitamins and minerals which play a vital part in any healthy diet. Gradually we are moving away from the idea that they are mere garnishes—an idea which started in commercial kitchens when the vegetable cook occupied the lowest position and was required only to clean the vegetables, put them in a large pot with plenty of water and cook them slowly to death. Vegetables are now recognised as a food and increasingly different varieties are being used in the home.

Undoubtedly the growth of the freezer market and the range of vegetables in Freezer Food Centres has been a major factor in making the public aware of the choice available.

Most vegetable seasons are very short and rarely do the more unusual vegetables become cheap. Those imported from abroad cannot possibly be fresh on arrival at the British market.

By freezing, vegetables are held in prime condition however arduous a journey they have to travel. They are available at all times of the year, which keeps prices stable, and they usually compare well in price to their fresh equivalents especially when out of season. On many imported vegetables such as peppers, broccoli and asparagus, a 25 to 50 per cent saving can be made throughout the year.

As I said in Part Two, to assume that frozen vegetables are better than fresh would be ridiculous, but the fact is that very few fresh vegetables—those one day old or less—are available unless they come from our own gardens. So as a substitute frozen vegetables are preferable to the fresh ones available in the High Street.

Whether the eating quality is affected by freezing is very much a matter of personal taste, but if cooked correctly there could be little cause for complaint.

Any nutrient content is lost not in the freezing process, overcooking is usually responsible for that.

Buying Vegetables

Vegetables are sold in a variety of pack sizes from 4 oz packs to 20 lb boxes. For the freezer owner 2–5 lb packs are usually the most economical. 20 lb boxes are certainly good value for money if the family prefers one particular vegetable, but to buy more than two 20 lb boxes would be over-stocking. Before investing in large economy packs check the quality—sometimes they contain a lower-grade product than the smaller packs. For your purposes that may be acceptable considering the substantial saving to be made. However, it would be wise to bear in mind that economy packs are not necessarily best quality.

As we eat vegetables almost every day, variety is perhaps of most importance. We have been very creative with potatoes over the years, serving them chipped, boiled, baked and creamed, to name but a few variations. But other vegetables are not so versatile and therefore different varieties should be bought. Frozen vegetables have cut out the monotony of eating in season, but this shouldn't be abandoned completely. Frozen vegetables

Many types of vegetable are suitable for freezing

certainly have many advantages, but they should never be used as a complete substitute for fresh ones. In season fresh may be fractionally cheaper and to some preferable in taste. A new saying could be coined: "Eat fresh in season and frozen out of season." Use fresh and frozen to complement each other and buy frozen vegetables when the fresh varieties are not available.

The average family eats approximately 1 lb of vegetables per day—that is a monthly total of about 30 lb. Perhaps one-third of these will be fresh—leaving 15 to 20 lb to be stored in the freezer. Certainly buy larger quantities of the vegetables most popular with the family but leave room for two or three small packs of more unusual vegetables and add a new dimension to the family diet. Not all need be served as a mere vegetable; some such as stuffed peppers or corn on the cob can be served as meals in themselves, while others such as celery or mushrooms can enhance the stews and casseroles.

The quality of frozen vegetables varies more than one might imagine—so check quality against price before buying: it's worth remembering "you get what you pay for".

Buying Guide

Peas Broad beans Sprouts French beans Cauliflower Carrots Spinach Swede

---

In season it is possible to buy these vegetables cheaper fresh than frozen—but check quality carefully. Also assess the amount of wastage incurred in preparation before comparing costs.

Broccoli Corn on the cob Haricots verts Asparagus Peppers Courgettes Petits pois

---

Seldom is it possible to buy these fresh in the local High Street as their season in Britain is very short. They are certainly to be recommended if available fresh but may be more expensive than frozen.

Frozen peppers, asparagus and courgettes are particularly good buys and certainly retain the individual flavour of the fresh product.

Mixed vegetables Corn kernels Artichoke hearts

Fresh equivalents are not available and so frozen must be compared with tinned. Frozen are perhaps a little more expensive, but the quality is definitely superior. When checking mixed vegetable packs ensure the contents are the same—obviously cheaper packs contain cheaper vegetables.

Potato chips and onions

By no stretch of the imagination could it be said that these are economical buys—but they must be top of the list for convenience. Sliced onions are particularly useful especially when only a little is required (no tears, either). Both are worth the extra pence per pound in sheer convenience.

Stewpack Macedoine Celery Tomatoes Mushrooms

Stewpacks are available in various forms—some compare well, some don't. There isn't a particular saving here, but certainly convenience. Celery and mushrooms can be used as required with none of the waste that occurs with fresh. Frozen tomatoes are not suitable for the salad but vastly better than tinned for all other uses.

Cooking Vegetables

It is a British obsession to overcook vegetables. Perhaps this was done for good reason in the past, either because of hygiene or to ensure absolute tenderness, but today it is completely unnecessary and with frozen vegetables overcooking gives poor results. If overcooked they lose much of their goodness, the texture is soggy and the flavour unrecognisable. The nutritive value is lost in the boiling water and the remains are eaten. Vegetables should be served crisp to appreciate their full flavour. As frozen vegetables have been through a blanching process, less cooking time is required than for fresh. To get the best results from frozen vegetables always cook from frozen for the minimum amount of time. Cooking instructions on packs vary from brand to brand, which is rather mysterious. The following instructions may alleviate the confusion.

The following method is for ½ lb of vegetables. For larger quantities increase the size of the saucepan and use proportionately more water.

---

1. Cover the bottom of the saucepan with an inch of water.

---

2. Bring to the boil, adding salt to taste.

---

3. Add the vegetables. Cook for the recommended number of minutes, drain well and serve immediately with a knob of butter.

A time span is given to allow for personal taste, but cooking for the minimum time retains the best flavour.

Asparagus	3–5 min	Cut green beans	3–5 min
Artichoke hearts	5–7 min	Diced carrots	3–5 min
Baby carrots	5–7 min	Fine whole beans	3–5 min
Broad beans	5–7 min	Haricots verts	3–5 min
Broccoli	5–7 min	Mixed vegetables	3–5 min
Brussels sprouts large	5–7 min	Peas	2–4 min
Brussels sprouts small	3–5 min	Petits pois	2–4 min
Cauliflower	5–7 min	Sliced bean pieces	2–4 min
Corn kernels	3–5 min	Spinach	3–5 min

Some vegetables require different methods of cooking. The following list offers some suggestions.

Courgettes—Sliced
Heat a small amount of butter or oil in a pan and fry the courgettes until well browned on each side. Cook with onion and/or tomato for extra flavour. Add plenty of seasoning and herbs as required. Makes an excellent ingredient for ratatouille or for any casserole dish.

Corn on the Cob
Cook for 8 to 10 minutes in just enough boiling water to cover the cobs. Drain well and then fry in a little melted butter until slightly browned. Serve with plenty of butter.

Potato Chips
Deep-fat fry straight from the freezer. Ensure the fat is hot before introducing the chips and cook for 3 to 5 minutes until golden brown. Drain well and serve immediately.

Onions—Sliced
Fry in hot fat until well browned, or use for stews and casseroles straight from the freezer.

Peas cooked from frozen need only 2 to 4 minutes in boiling salted water

Onions—Whole

Cover the onions with boiling water and simmer for 10 to 15 minutes. Drain and serve with a sauce. Alternatively braise or roast in the oven for approximately 45 minutes. They are perhaps best used in slow-cooked meat dishes and stews.

Stewpack, Celery, Macedoine

All three can be cooked and served as a vegetable, but are better used for soups, stews and casseroles. Use straight from the freezer and add to the dish 45 minutes before the end of cooking time.

Peppers

A versatile vegetable which can be used in all meat dishes to give a distinctive flavour. Also a main ingredient in ratatouille.

Vegetables can be stored in the freezer for varying periods from 1 month to 12 months

Whole peppers can be stuffed with a variety of fillings and served as a snack or meal. Remove the stalk and seeds—with care this can be done while still frozen—and cook in boiling water for 6 to 7 minutes. Drain, fill with stuffing and finish off in the oven.

Swede

Swede is the exception to the rule—it should be overcooked. Boil for 10 to 15 minutes in sufficient water to prevent burning. Mash with a fork and serve with plenty of butter and freshly ground black pepper.

Mushrooms

Blanching concentrates the flavour of frozen mushrooms—so less are required for flavouring stews or soups. They can be served fried or grilled, but ensure they are good quality—otherwise they can be a little tough.

Tomatoes

Excellent for grilling and frying. Use also for flavouring. They can be skinned easily by holding under the hot-water tap for a few seconds.

5. FRUIT

Many of the suggestions put forward in the last section on vegetables can be applied to fruit.

However, most fruits have a high water content and on thawing tend to go rather limp. The flavour is excellent, but the texture is changed somewhat. For the best results frozen fruit should be used only for cooking or incorporating into home-made desserts. Tinned fruit on the whole is cheaper and it is a matter of taste which is preferred, but without doubt frozen fruit is nearer to the true fresh flavour.

Pack sizes are generally small—1 lb or 2 lb. Free-flow packs are advisable as all the contents may not be required at one time. Those containing syrup are by necessity in block form but can be split if only a half-pack is required.

Check before buying if the fruit has been frozen in sugar—most are packed in their raw state so that they can be used as required. Freezing in syrup is necessary for some fruits to preserve the colour and texture. Fruits frozen down in their raw state retain a better flavour: sugar tends to give them a "tinned fruit" taste.

Frozen fruits are best thawed slowly in the refrigerator and used as required. If they are to be stewed or included in a dessert to be cooked, they should be used from frozen. Add sugar to taste in all cases.

For eating raw, fresh fruit has no substitute, but for cooking purposes or for making desserts frozen fruits are excellent. They are available all year round and many creative desserts can be made without seasonal restrictions.

Buying Fruit

Apples
These are usually cooking apples, sliced or chopped. As they are priced at edible weight they usually compare well with fresh. They often make a better apple pie or flan than fresh apples—the blanching process makes then fluffier.

Blackberries, Blackcurrants, Gooseberries, Plums and Rhubarb
These fruits fresh are usually cheaper in mid-season, but only gooseberries and rhubarb ever seem to be around long enough to take advantage of it. They all freeze well, but should be cooked in some form before eating.

Many varieties of fruit can be frozen when in season

Apricots, Mandarins, Pears, Pineapples, Raspberries, Cherries, Grapefruit and Peaches

These fruits tend to keep their shape well on thawing and are excellent for use in cold desserts, mousse or ice-cream. Mandarins and grapefruit sections can be rather bitter and should be coated in sugar before thawing. Raspberries are the all-time favourite frozen fruit and with good reason.

They all usually compare well for price with fresh. Mandarins, peaches and grapefruit are in season a better buy fresh.

Fruit Salad and Melon Balls

Both are frozen in syrup, but retain their flavour well. Melon balls are excellent as a starter, particularly if entertaining. Fruit salad may initially seem expensive, but the price of a home-made one would be very similar. Both are convenience foods, but are occasionally good buys.

Fruits can be made into prepared dishes before freezing

Lemons
Usually sold as lemon slices. Not a particularly economical buy, but very useful, particularly when an odd slice is required. Use for garnishes and cocktails but not for juice, as it works out too expensive.

Strawberries
To first-time buyers usually a disappointment—they look so appetising when frozen but on thawing go to a mush. Excellent flavour for use in jellies, flans, mousse, etc., but certainly not to be served on their own with cream.

Fruit juices
Most fruit juices are of superb quality and reasonably priced compared to tinned and certainly cheaper than home made. They are concentrated and so take up very little freezer space. As well as an excellent accompaniment to breakfast, these juices have a variety of other uses particularly in flavouring sauces, fruit desserts and party punch.

6. FISH

Introduction

For many years, fish has been a relatively inexpensive source of protein, and yet unfortunately it has never enjoyed mass popularity. Lately, however, perhaps because of soaring meat prices, the versatility of fish has been recognised, and it now features more prominently in the normal family diet. Fish prices are regulated not necessarily by demand, as with many foodstuffs, but by the supply available. Many of our waters have been overfished in the past, and we are now experiencing a shortage of the more popular varieties—subsequently prices go up. But the situation is by no means critical: there are many varieties still available at low prices, and it is only a matter of time before we are eating these fish on a regular basis. Most of us at some time have enjoyed coley, rock salmon, saithe or huss from the local fish-and-chip shop, but we buy it at the fishmonger's for pets only. If cod and other well-known varieties remain at inflated prices, no doubt coley will become increasingly popular.

The price of fish is no indication of its nutritive value or flavour—fashion and availability seem to rule in this respect. The price of salmon is three or four times that of herring, and yet the nutritive value of the latter is higher. Flavour, of course, is very much a personal choice, but in my view that of a kipper would be hard to beat. Both herring and salmon are very versatile and can be made into nutritious dishes. The nutritive value in all fish is high, and should be exploited to its full potential.

More than half the fish sold in Britain is frozen, or has been frozen—quite a staggering revelation. Why should this be so? Fish is at its best only when eaten fresh—it deteriorates fast once landed, and must be eaten within 48 hours to be appreciated to the full. By freezing at sea or at harbour factories, fish is preserved in prime condition for transportation to shops and markets. It is then sold either frozen as in the supermarket or fresh on the fishmonger's slab. From the consumer's point of view, it is surely better to buy the fish while still frozen, to be used as required, rather than the "fresh", which is probably deteriorating at a reasonably fast rate. Not all fish has been frozen, but it is wise to check before buying, and, if it is fresh, check *how* fresh.

Fish bought at the harbour or from fisherman friends can be guaranteed fresh—but for most of us that will remain a pipe-

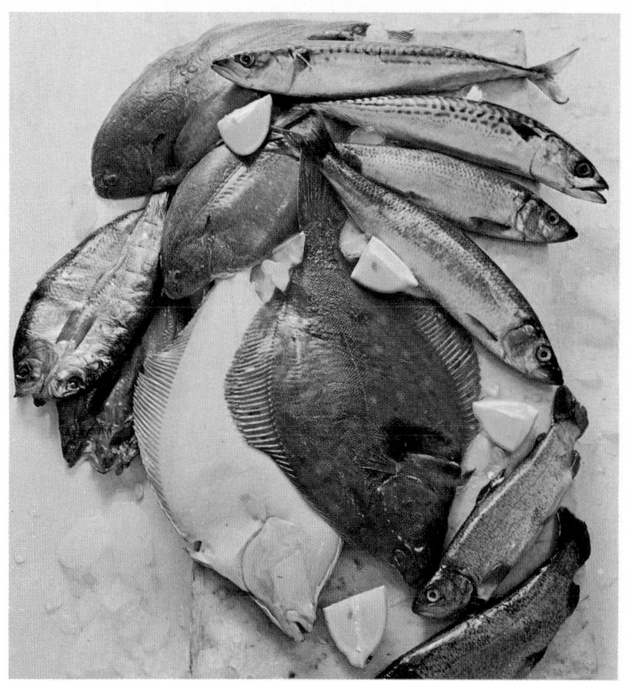

All types of fish can be available throughout the year when stored in a freezer

dream.

Fish, then, is one of the best frozen products. Freezer Food Centres hold a wide variety of fish which is not regulated by the time of the year. Prices are very competitive in comparison with those in the High Street, and substantial savings are possible.

As we are surrounded by the sea, one would expect us to have a healthy appetite for fish, but still for many it is a meal to be enjoyed only occasionally. Perhaps as the Freezer Food Centres offer more variety, the housewife may appreciate the versatility of fish and get away from serving it only fried or grilled.

Buying Fish

There are so many varieties of fish that to discuss each in detail would need a book in itself. The following are the selection most commonly available frozen.

Bake frozen fish in foil to retain maximum flavour

White Fish

Plaice, Dover Sole, Lemon Sole, Coley, Cod, Halibut, Haddock.

All white fish is suitable for frying or grilling, but is used also for poaching, steaming and baking. The flavour of white fish is rather delicate, and, although it is pleasant served on its own occasionally, it is certainly improved with a sauce or incorporated into one of the numerous varieties of fish dishes.

White fish is cut and processed in various ways, increasing the choice of purchase. Cod, coley and haddock are available whole or filleted in steaks and cutlets, and are used also for fish fingers and fishcakes. Halibut is usually sold in cutlet form, although fillets are available.

Small whole plaice are available as well as breaded and unbreaded fillets. Lemon and Dover sole are sold mainly on the bone, and should be eaten as such to appreciate their full flavour. Prices vary, but halibut and Dover sole are rarely a budget buy, whereas coley, cod and haddock are.

Oily Fish

Herrings, Whitebait, Mackerel, Sprats.

Other common oily fish are sold almost exclusively in tins, e.g. tuna, anchovies, sardines. Oily fish are very nutritious as well as having a good flavour. They can be cooked in a variety of ways, but are increasingly used as starters, or just on their own as a fish course. As these are small fish they are invariably sold on the bone.

Smoked Fish

Smoking certainly enhances the flavour of some fish. It can be applied to all, but most commonly available are kippers, haddock, salmon, trout and mackerel.

Traditional breakfast dishes, kippers and haddock, are available on or off the bone. Boil-in-the-bag varieties enjoy increased popularity and, although a little more expensive, are very convenient.

Smoked salmon, trout and eel are still luxury buys, but as an *hors d'oeuvre* or to complement a cold table are worth a little purse stretching.

Shellfish

The two main groups of shellfish (crustaceans with jointed shells—lobsters, crawfish, crabs, prawns and shrimps—and molluscs with hinged shells—oysters, scallops and mussels) provide a valuable source of protein but are, in the main, expensive.

With frozen shrimps and prawns there is a substantial price saving of up to 50 per cent compared to those bought fresh or tinned. They are ready-cooked and sold either shelled or unshelled. Scampi are also a good buy. Breaded scampi is popular, but unbreaded is more versatile. Scallops are similar to scampi and are generally a little cheaper. Breaded scampi and scallops should be deep-fat fried and served with a selection of vegetables.

Crab, lobster and crawfish are sold both with and without the shell. The meat is usually cooked and can be used for pâté, with salad or for special occasion dishes.

Shellfish delicacies are available on demand

Freshwater Fish

The two types most generally available are salmon and rainbow trout. World-famous Scottish salmon is available, but those from Japan and Canada are good and much cheaper. If buying a whole salmon make sure that you have adequate cooking facilities—otherwise steaks are easier to handle. Salmon can be used in a variety of prepared dishes or just poached or steamed in the normal way.

To preserve the delicate flavour of rainbow trout, grill and baste frequently with melted butter. Left-overs? Try pâté or mousse.

Pack sizes range from 1 lb to 5 lb, the larger packs usually being the most economical. Don't overstock on fish as its storage life is somewhat limited, particularly that of shellfish, and so the maximum buy should be stock for two months. As the range is so vast, choose carefully and analyse prices. A 5 lb pack of breaded fish, for example, is very convenient but only if the family eats it regularly. Perhaps a pack unbreaded could be more versatile, and occasionally the fish could be coated at home if required. A 4 lb pack of salmon steaks may seem an extravagant buy, but is it? Compared to a fillet steak, it is relatively inexpensive.

The wide range of fish can be rather confusing to those unfamiliar with it and so shop around carefully, but don't overlook lesser-known varieties. Fish and a good recipe book can produce surprisingly professional results.

Cooking Fish

Frozen fish should be cooked from frozen, where possible. As with meat, if thawed, many of the valuable juices are lost and the fish can be dry and less appetising.

For grilling and frying, add a third to the normal cooking time, on a reduced heat. Baste well to prevent excessive drying on the surface.

If the fish is to be incorporated into another dish, thawing may be necessary. This should be done overnight in the refrigerator. Any "drip" should be used in an accompanying sauce or in the dish itself.

Where possible, use fish in interesting ways and be creative. A previously uninterested family could become fish lovers with the right kind of persuasion, and this could help the food budget.

Salmon need no longer be a luxury item

7. BACON AND GAMMON

It was thought at one time that bacon was not an ideal subject for freezing. In fact it freezes well. Its storage life is two to three months, which is not restrictive since there is not a sufficient price advantage to prompt you to buy in more than a two- or three-month supply. However, it is a comfort to know it is ever present. There are three main varieties of bacon available, the taste differing according to the curing process used. Brine curing gives greenback bacon, smoked bacon is produced by hanging the loin over smouldering wood, and the sweetcure taste is created by curing in a sugar solution rather than salt, which has a tenderising effect on the bacon. Of the three, greenback and sweetcure are the best for freezing.

Bacon rashers are sold usually in 2 lb or 3 lb packs, but ensure that 1 lb or ½ lb units are easily separated, preferably vacuum packed. If the pack has been frozen in block, it is impossible to separate without spoiling some of the rashers. Don't overstock as a rancid flavour is detectable after prolonged storage. This is in no way harmful, but the taste can be unpleasant.

Once thawed, store in the refrigerator and treat as fresh. It is unwise to store bacon for more than three days at refrigerator temperature, as off flavours develop quite quickly.

A gammon joint is always a good stand-by for entertaining

There is a variety of bacon joints—knuckles, collar and forehock being the most common. These economical boiling joints are versatile and can be served hot or cold as preferred.

To get the best results, the preparation is rather lengthy. Short cuts can be made but are not recommended unless time is limited. Treat all types of bacon joints in the same way, but remember if you are tempted by a really big joint that you must have a pan large enough to accommodate it.

Soak the frozen joint for 24 hours in cold water. Drain and re-cover with fresh water. Add a *bouquet garni* together with a few ounces of sliced onion and bring to the boil. Simmer gently for 35 to 45 minutes per pound depending on the cut. Remove the joint from the pan and strip off the skin. If the joint is to be served cold, replace in the water to cool as this prevents it drying out. If it is to be served hot, score the fat in a criss-cross pattern and crisp up in a hot oven for 10 to 15 minutes.

If there isn't time to soak the joint for 24 hours place the frozen joint in a pan, cover with cold water and bring slowly to the boil. (This will take some time as the joint will have to thaw.) Pour off this water and replace with fresh and continue as with the previous method.

The same procedure applies to gammon joints, although the boiling time will be shortened to 20 to 25 minutes per pound. It is difficult to be exact on cooking times for bacon and gammon joints as the cut and quality are the determining factors.

Bacon and gammon are worthwhile freezer buys—select for quality rather than price to obtain best value for money.

8. CAKES, PASTRIES AND BREAD

Unfortunately home baking in the true sense is practised in very few homes today. Most of us do not have the time or perhaps the inclination to make our own bread, only a few cakes and pastries occasionally. Never before has so much bought baked produce been eaten in Britain.

If our confectionery compared with that of the Continental *patisserie* one would have little cause to comment, but so many of the cakes, pastries and loaves that we eat are little better than moulded plastic or sliced cardboard. We seem to have cultivated a taste for synthetic, over-sweetened, third-rate desserts and have forgotten the superb taste of home baking.

Perhaps the freezer may help to rectify things in the future. Time is the all-important factor and a weekly baking session for most busy housewives is out of the question, but with a freezer baking can be done once a month and the results stored until required. I do believe that this is where the freezer can save money and give a lot of satisfaction. It wouldn't, however, be feasible to suggest that every morsel should be home made—obviously a certain amount will be bought ready made. The range available at Freezer Food Centres is vast. But it must be remembered that frozen confectionery is a convenience food and thus is not particularly economical, although some items are better value than others.

Cakes

Don't be tempted to buy decorative packaging—it's what's inside that counts. Price seems to have little relation to quality, and so rely on recommendation if possible.

Yeast-based mixtures—Danish pastries, doughnuts, etc.—freeze well and these are possibly the ones seldom made at home. Cheesecake in its various forms is always a good buy. For those unfamiliar with cheesecake it probably seems an extravagance, but as the mixture is so rich only a small portion is required—in fact, portion for portion, it is no more expensive than many other cakes. It has an unusual flavour—it is perhaps even an acquired taste—but for home entertaining it is becoming increasingly popular.

Gateaux on the whole I cannot recommend as a good buy. The products themselves are usually average, but there are problems in transporting and storing a decorated cake which to my mind make the whole venture a waste of time. Although frozen, they are still liable to damage if not treated with care. Gateaux are better bought fresh from the baker or made at home. A simple method of producing a home-made gateau is by decorating a bought plain sponge cake. Dairy cream and jam sandwich sponges are available in boxes of six to eight, giving a considerable price saving. With a selection of fruit, cream, nuts, etc., and a little creativity, a home-made gateau can be produced for half the price of a bought one. Plain sponges are easy to store and, although splendid eaten plain, can form the base of an instant gateau for the unexpected visitors. Cakes require varying lengths of time to thaw depending on the density. Unlike some other products a slow thaw is not essential. Room temperatures give quicker results. Unused portions can be refrozen.

Sponge cakes, Danish pastries, cheesecake, etc.	2–4 hours
Fruit cake and other dense mixtures	6–8 hours

For the best results, eat as soon as possible after thawing.

Personal taste will weigh heavily when purchasing cakes, but in my view cheesecake, Danish pastries and plain cream sponges would be hard to beat.

That home-made flavour can be stored in the freezer

Pastry and Pastry Products

Pastry freezes very well but as with cakes the products themselves vary in quality. The best buys are undoubtedly packs of puff and shortcrust pastry. Choose a reputable brand name to ensure top quality and check that the pastry is packed in ½ lb or 1 lb units for easy use. The difference between bought and home-made pastry is negligible: the cost is perhaps a little more, but the time saved surely warrants the difference, particularly in the case of puff pastry for which few of us have mastered the art. Pastry can be refrozen, which is a big advantage. So, having thawed and been shaped, it can be refrozen either cooked or uncooked without affecting the quality.

Pastry cases of various kinds are also useful, in particular

Frozen Danish pastries make an excellent supplement to home baking

ready-to-cook vol-au-vent cases—available in varying sizes. Home-made ones are fiddling and often misshapen, whereas the frozen ones only require baking and are 100 per cent reliable. Stuffed with a variety of interesting fillings they can be served as snacks or as a main meal, and the party buffet wouldn't be complete without them.

Bought pies and their many variations are generally of reasonable quality and certainly one or two in the freezer are a boon for the odd occasion. They should always be cooked from frozen. Exact times and temperatures are usually given on the box.

Good-quality pastry products are a useful stand-by, but it is relatively simple to produce home-made pastries from the products in the freezer. Frozen pastry together with a choice from the variety of frozen fruits can be assembled to make home-made desserts in a short space of time, and although not genuinely home made they do have that unique flavour so that the family will never detect the difference. There is perhaps a little difference in price, but the eating quality is better.

Bread

No freezer is complete without a good stock of bread. Bread freezes extremely well and it has been hinted that freezing actually

Frozen pastry can be made up into a variety of dishes and refrozen

improves its quality. The reasons cannot be explained, but the fact remains.

To obtain fresh bread it must be purchased daily, but in the freezer it will retain that freshness for up to three months. It would be ridiculous to stock up with three months' supply of bread at a time, as that would fill half the freezer. So stock weekly or fortnightly depending on the amount consumed. For those who eat very little bread, wastage is reduced to a minimum as only the required amount needs to be used at a time.

Home-made bread has no rival and although time-consuming it is relatively easy to make. It certainly is becoming a fashionable occupation and if the bread is made in bulk quantities the freezer will keep it fresh for weeks. Frozen bread dough is now available and could be a good start for those who are new to bread-making. It makes high-quality bread at a reasonable cost.

The alternative is to find a good bakery, preferably where the bread is baked on the premises, and buy a selection to store.

No freezer should be without a selection of bread

Sliced breads are useful for toast and sandwiches, the main advantage being that only the required number of slices need be removed at one time. For toast do not thaw the bread but grill from frozen: it gives surprisingly good results.

The storage life of bread varies. Crusty loaves, French bread, Vienna sticks, etc., do become a little dry if stored too long, and, on thawing, the crust tends to "shell off". Storage life should be kept to a minimum.

Soft bread and rolls seem to have an eternal storage life but to keep large supplies would be an uneconomical use of freezer space. Thaw unwrapped at room temperature for one to three hours. Rewrap or place in the bread tin.

No doubt frozen bread in various forms and shapes will be introduced before long and if it proves to be price-saving then it will be a worthwhile buy. However, for the moment, the baker or the supermarket can provide a much wider range to choose from. The most important factor is to buy "fresh".

9. DAIRY PRODUCE

The benefits of freezing dairy produce are not obvious as all the commodities under this general heading are available all the year round and not subject to seasonal price fluctuations, although in the past few years prices have fluctuated for various other reasons. With the exception of ice-cream, few dairy products are sold in bulk quantities in Freezer Food Centres, although, of course, they are available at many cash-and-carry warehouses, but unless the price saving is significant they are not a worthwhile buy.

It is, however, very useful to store small quantities of milk, butter, eggs and cheese for emergency use, or to cut down on supermarket shopping.

Ice-cream

Ice-cream sales have soared with the arrival of the home freezer. It is no longer just something to lick in the garden on a warm summer's day. To the home-freezer owner, ice-cream represents a nutritious and very acceptable dessert.

Quality varies from brand to brand, but on the whole the supermarket's own-label ice-creams are as good as the well-known brands available at a higher price. Personal taste will obviously affect this judgement and one can usually count on the children in the family being the connoisseurs.

Buying ice-cream in quantity is economical. The difference between a gallon container and its equivalent in small 4 oz packs can mean a difference of 30 to 40 per cent. While it must be borne in mind that with ice-cream, perhaps more than with any other product, consumption increases with availability, nevertheless gallon or half-gallon containers are a good buy. Approximately 60 scoops of ice-cream are available from a gallon container; so if previously the local ice-cream van has provided treats for the children the saving is very substantial. Wafers and cornets are also available in large quantities, but don't buy too many at one time unless they can be stored in airtight tins as they lose their crispness after a while. Buying large quantities does restrict variety—so choose one or two large containers of popular flavours, but add to this a few small packs for variety.

Lollies, mousse and many other frozen desserts are also available, and again the saving is worth while.

Cream

Whipping and clotted cream freeze well and small quantities

Ice-cream is a particularly good buy and is a nutritious family food

can be stored for convenience. Single and double are not particularly successful as they separate on thawing. However, if they are to be used in cooked dishes, they are satisfactory, but it would not be wise to buy and store for this reason. Left-overs can be utilised in this way rather than being wasted.

Butter

It is quite satisfactory to freeze butter, but unless it can be obtained at a very much reduced price it is not worth storing in bulk. Unsalted butter freezes better than salted, but either can be kept for a considerable time.

It is useful to store a ½ lb or 1 lb pack of butter, margarine and lard, just in case you run out, particularly at holiday times. Also when away on holiday it's nice to have these essential commodities in stock to come back to.

Cheese

All cheeses freeze well, but again it is an uneconomical use of freezer space to store in large quantities. Soft cheeses should be frozen only when mature so that on thawing they are ready to eat. Hard cheeses should be carefully wrapped to prevent excess drying. They will tend to be rather crumbly when thawed, but there will be no loss of flavour. After a party there is inevitably cheese left over—it is useful, therefore, to be able to store until the cheeseboard is required again.

A freezer helps to prepare a fresh cheese board and copes with the inevitable left-overs

Eggs

Eggs do not freeze particularly well, but it depends on how they are ultimately to be used. Eggs can't be frozen in their shells, as expansion causes them to crack. Each egg should be cracked and frozen separately—ice-cube trays are useful for this. Alternatively, several eggs can be beaten together and frozen in a container; the addition of a small quantity of salt or sugar prevents coagulation. These can then be used for various egg dishes, such as omelettes, or for cake making. Also egg whites and yolks can be frozen separately for use in sauces, icings or meringues. Freeze left-over quantities of egg which can be used for glazing or coating.

Ice-cream is certainly a worthwhile buy, but the storage of other dairy produce should be restricted to small quantities and left-overs. The fact that one has the facility to save left-overs rather than throwing them away is an economy in itself and should be used to the full.

Part Four
Home Freezing

1. INTRODUCTION

The case for and against home freezing was discussed in Part Two, where my feelings will have been made clear. It might be fun to freeze down some raw materials, particularly those from the kitchen garden, but it can hardly be described as economical. A much stronger case can be made for the freezing of ready-prepared or part-prepared dishes. This can be both fun and highly economical. Many ready-prepared meals are available from Freezer Food Centres and are admirable in their way, but they are seldom as good as Mum makes.

I believe that the raw materials are in the main better purchased frozen. Certainly they will have been frozen at their peak. As illustrated on page 26 the cost of freezing peas bought fresh shows no saving over the best grades of frozen peas even without taking one's time into account. Furthermore, Birds Eye and others go to great lengths to freeze their peas within two hours of harvesting. Those bought from the greengrocer will be more like two days old and, believe me, Birds Eye don't hurry like that for fun.

Nevertheless in this age of rising prices inevitably many are going to invest a lot of time in growing produce for the freezer. For the first time a means of preservation is available which will hold the food in its natural state until required.

Most of this part of the book is devoted to the garden and to the procedure for freezing the harvest to obtain successful results. Timing and accurate preparation are important. The sections on meat and fish are comparatively short as their preparation is simple. But don't make the commonest mistake of all and try to freeze down too much at one time. Many purchase freezers with the sole intention of storing meat bought at wholesale prices, then proceed to try and freeze down a cabinet full of "fresh warm" meat. The capabilities of the machine must be borne in mind otherwise the produce suffers.

The section on prepared foods is by necessity rather general. This was never intended as a recipe book and in order to pass on detailed information about the preparation and freezing of home-made dishes it would have required their inclusion. So although short I hope it serves as a guideline by which to adapt your own recipes.

2. WHY IT IS IMPORTANT TO FREEZE FOOD QUICKLY

Freezing is by far the best method of preservation as it changes neither the structure nor the flavour of the food to any significant degree. The nutrient content remains the same and the eating quality is identical to that of fresh food in most cases.

But this standard of preservation is achieved only if the food is frozen down quickly. All foods have an interior cell structure which can be destroyed if freezing is done too slowly. As food freezes, ice crystals form within the structure—the faster the food is frozen, the smaller the crystal formation, keeping the interior structure intact as far as possible. Slow freezing produces large ice crystals which destroy the structure both internally and externally.

As pointed out in Part Three, foods with a high water content such as strawberries, cucumbers, etc., never freeze very successfully as even the tiniest crystal formation breaks down the very delicate structure. On thawing they go rather soft and mushy but can be utilised in dishes or for cooking as the flavour is unchanged.

Food frozen slowly is disappointing in texture and flavour, but of more importance is the loss of nutrients when defrosting or cooking. The correct use of the fast-freeze device is important.

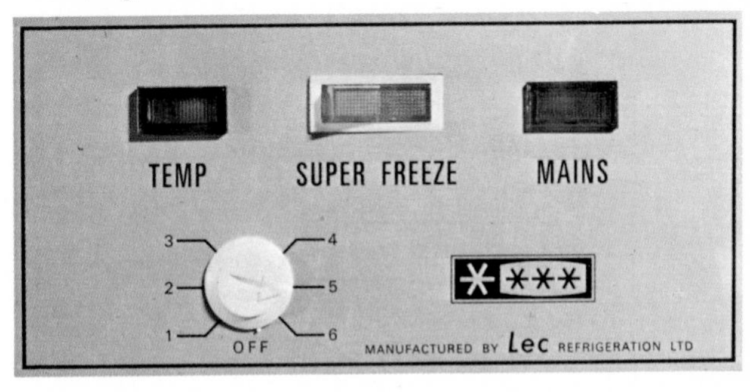

A control panel featuring the new freezer symbol

3. USE OF THE FAST-FREEZE COMPARTMENT

The "fast-freeze compartment" has assumed many titles—e.g. "super-freeze", "automatic freeze", "extra-cold", etc. Whatever creative name is thought up by the manufacturer, this compartment is the place where fresh food is placed for freezing.

The action of switching on to fast-freeze overrides the thermostat, keeping the motor running continuously and pulling down the temperature in the freezer, enabling small quantities of fresh food to be introduced. Most machines will achieve their minimum temperature within a two-hour period, and so switch on to fast-freeze about two hours before introducing the food.

The performance of freezers varies considerably, and until you know the lowest temperature which can be achieved by your freezer it is difficult to estimate the amount which can be frozen at any one time.

All freezers are capable of freezing 10 per cent of their loading capacity, which for the majority of people is more than ample. This means that in a 12 cu ft chest freezer 24 lb of fresh food can be frozen in a 24-hour period. But for those who wish to be more specific it can be worked out quite simply with the aid of a thermometer. The following will serve as a guide.

1. Switch on to fast-freeze and take a reading on the thermometer after two or three hours.

2. The temperature reading should be between −28° and −35°C (−18° and −30°F). This will be dependent on the performance of the machine and the ambient temperature.

3. Calculate the overall loading capacity of the freezer. This is done by multiplying the net capacity by the amount which can be stored per cubic foot. For example:

   (a) *12 cu ft chest freezer*
   12 cu ft × 20 lb per cu ft = 240 lb maximum loading capacity

   (b) *14 cu ft upright freezer*
   14 cu ft × 15 lb per cu ft = 210 lb maximum loading capacity

4. The minimum temperature reading will fall into one of three categories:

   (a) −28° to −30°C (−18° to −22°F) temperature range. Allows 10% of maximum loading capacity to be frozen in 24 hours.

   (b) −30° to −32°C (−22° to −26°F) temperature range. Allows 15% of maximum loading capacity to be frozen in 24 hours.

   (c) −32° to −35°C (−26° to 30°F) temperature

range. Allows 20% of maximum loading capacity to be frozen in 24 hours.

5. Examples

(a) *12 cu ft chest freezer*—maximum capacity 240 lb
Minimum temperature −35°C (−30°F)
20% of 240 lb = 48 lb.

(b) *14 cu ft upright freezer*—maximum capacity 210 lb
Minimum temperature −32°C (−26°F)
15% of 210 lb = 31½ lb.

The more powerful machines, which are generally those having a fan-assisted motor, will freeze larger quantities of food, but in no way does this imply that these are superior machines. All freezers are built to do a similar job and few people would ever take advantage of this extra power unless involved in vast quantities of home freezing.

Fast-freeze compartments differ in size, and in some machines they are non-existent. Why? When switched to fast-freeze the whole of the cabinet lowers in temperature, not just the compartment. Most chest freezers do have a specific fast-freeze compartment, its purpose being to keep the "fresh warm" food away from that already frozen. Also the size of the compartment is a guide to the amount which can be loaded at any one time.

Some small chest freezers do not have such a compartment, as a divider would restrict the storage of large packages. In these instances "fresh warm" food should be stacked against the walls of the freezer and if possible away from the frozen food.

An upright machine is a series of compartments divided by shelving, and so with this type select a shelf either at the top or bottom of the machine. Clear the shelf of other produce and lay out on it the "fresh warm" food. If the shelves in the cabinet are not refrigerated then the food should be stacked against the walls.

Any kind of fast-freeze compartment should be used for normal storage when not freezing.

Obviously the quickest way to freeze the food is by having it in contact with the freezing elements, situated either in the shelves or in the walls of the cabinet. Therefore, when freezing large quantities, do not restrict yourself to the use of one shelf or compartment. It is better to spread the food around to get the maximum benefit from the cold surfaces. But care must be taken to ensure the safe preservation of food already stored. Restrict the quantities to be frozen where possible to ensure quick freezing.

The fast-freeze switch should not be left on for longer than 24 hours or otherwise the motor unit tends to get overheated. However, a low temperature having been maintained over this period, the process of returning to the thermostat regulator is not going to have very much effect on the temperature in the freezer for a good few hours. Once the cabinet is down to a low temperature, provided the conditions are favourable and the cabinet reasonably full, it will maintain that low temperature for another 8 to 12 hours. So, although the fast-freeze switch can remain on for only 24 hours, the freezing time is considerably more. Having switched off fast-freeze, do not disturb the food but leave for another 6 to 10 hours to complete the process.

Having completed one day's freezing, give the freezer a rest for 24 hours before doing more. Bear this in mind when buying and preparing. If the food has to wait three days before it can be frozen, it's not going to be in prime condition.

It's all very well discussing the intricacies of bulk freezing, but what about the small quantities, which are what most of us are going to be concerned with most of the time? How long will it take to freeze, for example, an odd loaf of bread or a couple of pies? Listed below is a general guide to allow you to make your own calculations at home.

Bacon, meat, poultry	2 hours per lb
Fish	2 hours per lb
Vegetables and fruit	1 hour per lb
Prepared meals	2 hours per lb
Bread and cakes	1 hour per lb
Pastries	2 hours per lb
Dairy produce and liquids	1 hour per lb
Example	
4 × 8 oz Rump steak	4 hours
2 × 1 lb Apple pies	4 hours
2 × ½ lb Sponge cakes	1 hour
2 × 1 lb Loaves	2 hours
Total	**11 hours**

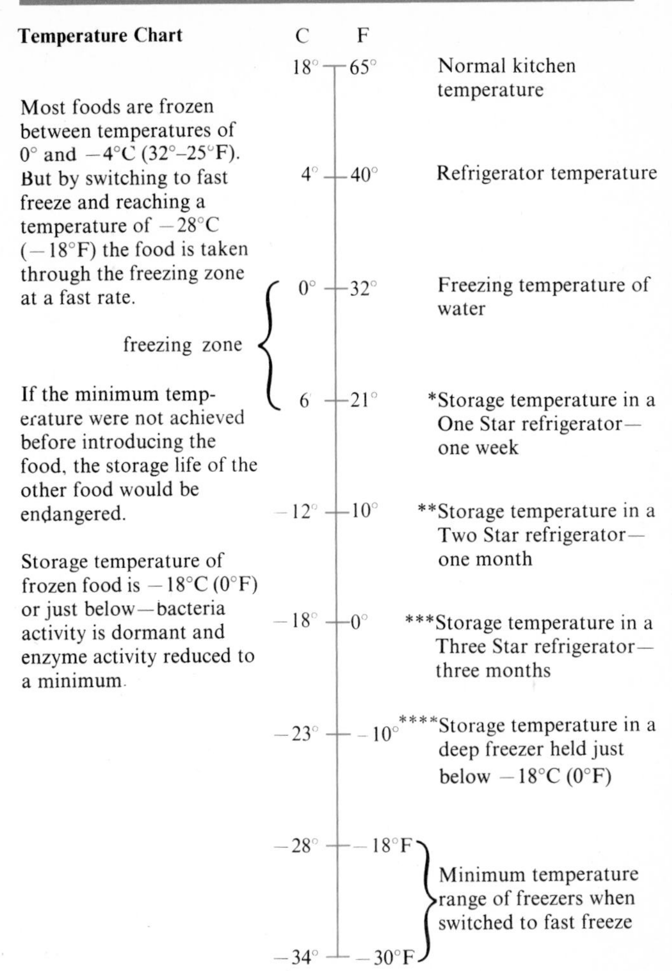

If the quantity to be frozen is 2 lb or less it is not necessary to switch on to fast-freeze as such a small amount will freeze quickly at storage temperature.

When freezing small quantities always stack against the refrigerated sides of the chest or on the shelf surface for maximum efficiency.

Once frozen, the food can be either left in the freezing compartment or transferred to a specified section in the freezer.

When the fast-freeze is switched off, the thermostat again takes over and regulates storage temperature in the freezer.

It is important to use the fast-freeze facility properly as it serves two purposes. Firstly it ensures that the foods are frozen down as quickly as possible, and secondly it is a protection for the foods already being stored. When "fresh warm" foods are introduced, the temperature goes up. If they were put in at storage temperature −18°C (0°F), the temperature could rise to −15° or −12°C (5° or 10°F), thus affecting the storage life of the food already in the cabinet.

4. FREEZING FRUIT AND VEGETABLES

Growing some fruit and vegetables for freezing at home can save you money, but this is dependent on making sensible use of the garden plot. By choosing the ideal varieties and freezing them in small amounts, you will be able to store a few pounds of your favourite vegetables and fruit—enough for a couple of months' supply.

Be warned—do not try to grow enough for the whole year. Not only will you have a full-time job gardening and freezing, but it will be an uneconomical use of your freezer space. Growing and freezing small amounts, about 5 to 10 lb of each vegetable or fruit, is fun—so don't overdo it and make it a chore. Ideally, choose six to eight varieties and aim for a 5 lb yield from each to freeze. The rest of the harvest should be eaten fresh.

Listed over the next few pages is information on the most popular fruits and vegetables, but in each area it is advisable to seek information from a local expert as climate and soil conditions may affect your final choice.

Vegetables recommended for home freezing	
Beetroot	Housewives' Choice
Broccoli	Any variety of Green, White or Purple Sprouting
Broad Beans	Harlington White White Windsor Masterpiece Green Giant Longpod Meteor White Eye Aquadulce Claudia

Brussels Sprouts	Noisette Irish Elegance Peer Gynt Darkcrop Early Button
Carrots	Early Nantes Champion Scarlet Horn Chantenay Amsterdam Forcing Cluseed New Model
Cauliflower	Improved Snowball Majestic
Corn on the Cob	Earliking John Innes First of All Kelvedon Glory Canada Cross
Courgettes	Early Gem
French Beans	Masterpiece The Prince Tendergreen Deuil Fin Precoce Phenix
Peas	Onward Thomas Laxton Early Onward Progress No. 9 Sweetness Kelvedon Triumph Dark Skinned Perfection Peter Pan Petits Pois Gradus Tall White Sugar Pea
Runner Beans	Scarlet Emperor Enorma 21 Kelvedon Wonder
Spinach	Perpetual Green Market

Fruits recommended for home freezing

Apples	Bramley Seedling Woolbrook Russet Oxford Friend Sowman's Seedling Corry's Wonder
Apricots	Moorpark
Blackberries	Himalaya Giant Ashton Cross Merton Early
Blackcurrants	Baldwin Boskoop Westwich Choice Seabrook's Black Wellington XXX
Cherries	Governor Wood Noir de Guben Circassian Montmorency
Gooseberries	Keepsake Careless Lancer Leveller
Peaches	Hale's Early (stone-free)
Pears	Burre Bedford Clapp's Favourite Gorham Louise Bonne Williams' Bon Cretien Doyenne du Comice
Plums	Shropshire Prune Damson Belle de Louvain Apricot Gage Jefferson's Gage
Raspberries	Malling Enterprise Exploit Jewel

	Promise Lloyd George Phyllis King September Wellington Norfolk Giant Early Red
Redcurrants	Laxten No. 1 Red Lake
Rhubarb	Timporley Early Champagne Linnaeus
Strawberries	Cambridge Vigour Cambridge Prizewinner Cambridge Rival Cambridge Favourite Templar Royal Sovereign
Tomatoes	Moneymaker Potentate

Fruits should be harvested in prime condition, either when fully ripe or just before.

Overripe or damaged fruit should be prepared and eaten fresh within a few days.

Sowing and Harvesting Vegetables

Beetroot
Small, tender young beets are ideal for freezing. The first sowings should be made in early April. Sow in rows 12 inches apart and thin the plants to 3 to 4 inches apart. Successive sowings may be made as the ground becomes available until early July. Larger beetroot should be sliced.

Broad Beans
February or March is the best time for sowing broad beans. Sow in double rows about 8 inches apart, leaving a space of 2½ feet between the double rows. The Masterpiece Green Giant Longpod variety is recommended because of its good colour after cooking.

If a smaller bean is preferred without a black eye, use a Meteor White Eye. Pick while still young or otherwise they can be tough.

Dwarf Beans

Most people in Britain grow the old flat-podded types, and, where this is preferred, use The Prince. Pick the pods young or otherwise they tend to be stringy. Sprite is highly recommended, too. The pods are pencil-like in shape and completely stringless at all stages of growth. Sow from the middle of April until June in rows 18 inches apart. If possible, pick when very young, so that they are small enough to freeze whole.

Variety of dwarf French bean ideal for freezing—The Prince

Runner Beans
Two varieties are ideal for the freezer—Kelvedon Wonder and Enorma 21. Kelvedon Wonder is a short-podded variety, early maturing and continuous cropping, whereas Enorma 21 produces pods which are frequently 18 to 21 inches in length and of first-class quality.

Most growers prefer to raise plants in trays in a cold frame or greenhouse. Sowing should be made on the first day of May for transplanting to the garden at the end of the month. This method avoids the risk of damage by May frosts.

There are various methods of cultivation. The plants may be grown up bean sticks, but, if these are difficult to obtain, erect two good stout poles, one at either end of the row, running between them a wire 6 inches from the ground, another across the middle and a third at the top at a height of 7 to 8 feet. Then run strings looped over the wires, from the top to the bottom, for the beans to climb up. Spacing should be 10 to 12 inches from plant to plant.

In a small garden, where you have to conserve space, erect a centre pole and run strings down in the shape of a bell tent, anchoring the strings with pegs. Sow two seeds alongside each peg and while the beans are growing cultivate a selected salad crop on the inside which will be cleared long before the beans close in. Harvest when young and tender, before the beans form inside the pod.

Broccoli
Sow the seed bed in the latter part of April, transplanting when the plants are large enough. Space carefully, allowing 20 inches between the rows and 18 inches between each plant. Harvest when head is firm.

Brussels Sprouts
Seeds should be sown in a seed bed outdoors at the end of March, or throughout April, transplanting in the open ground in early June.

Dig the ground early, to allow it to consolidate as sprouts like a firm anchorage. Only tiny button sprouts should be used for freezing, so harvesting must be well timed. The buttons should be firm and bright in colour.

Carrots
The perfect carrot for freezing is the shape and size of the middle finger. Like beetroot, young produce is the best, and the variety which is most suited is Champion Scarlet Horn. Seeds may be

sown in the open ground from the end of March until early July, in rows 1 foot apart, thinning the seedlings to 3 inches in the rows. Harvest when very small—baby carrots are best for freezing whole. Larger carrots should be sliced or diced.

Variety of brussels sprout ideal for freezing—Early Button

Variety of carrot ideal for freezing—Champion Scarlet Horn

Cauliflower

Many varieties respond well to freezing, but one which is probably more popular is Improved Snowball. It is necessary to raise the plants in a cold frame in March, transferring them to open ground as soon as possible, allowing 18 inches between the rows and the plants. Cut when the heads are still small and creamy white in colour, and freeze in florets.

Corn on the Cob

Although not difficult to grow, to avoid frost do not sow before the beginning of May. Place in rows 3 feet apart, spacing the plants at 15 inch intervals in the rows.

Use an early-maturing variety such as First of All. Well-grown plants should carry three cobs which will be ready for gathering during September.

It is possible that the beginner might find it difficult to know when the cob is ready for gathering as it is encased in a sheath. The tassel at the end of each cob changes colour as it matures, and when this begins to turn a mustard colour the cob is ready to harvest.

Courgettes

Best variety is Early Gem. It grows well in the greenhouse, or it can be sown under glass in April and transplanted in June. Courgettes can be frozen either whole or sliced, and so harvesting time should depend on how they are to be frozen.

Peas

It is important for the amateur grower to be fully acquainted with the various characteristics of each variety. The difficulty (and the cost) of procuring pea sticks is a case in point, and, where this problem arises, then dwarf-growing varieties such as Onward, Early Onward and Progress No. 9 are recommended. All three have a maximum height of 2 feet when fully developed, and are good for the small garden where space is limited. The rows can be sown much closer together than for the tall-growing varieties. Seeds should be placed 18 inches to 2 feet from row to row. If it is possible to grow a 3 foot variety, then Sweetness is recommended. It is thin shelled and matures very early.

Undoubtedly, the best variety of all in the tall-growing section (4½ feet) is Show Perfection—outstanding as it has all the characteristics required not only for growing but also for freezing. Heavy cropping and double podded, it has the thinnest pod of

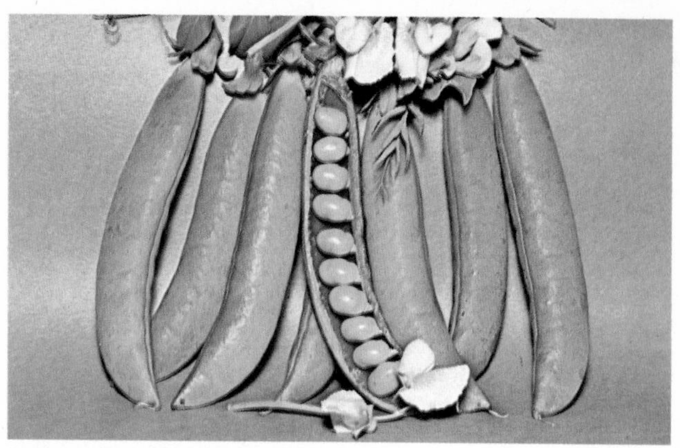

Variety of pea ideal for freezing—Sweetness

all known varieties and contains 12 peas in each. It achieves and maintains a dark-green-coloured pea before and after freezing.

Reference has already been made to the difficulty in obtaining pea sticks, but this should not deter any grower from producing this variety. As a substitute for sticks, use wire netting or a deep fish netting.

When the crop is in full flower, take the precaution of controlling pea maggot. Many sprays are available. This will save you the nasty job of picking out maggot-infested peas when shelling.

The most reliable date for general sowings is the first week in April for June and July picking. If you have the ground available, later sowings, using dwarf varieties, may be made until June.

Harvest when pods look full. Test one or two first—they should burst open easily with a little pressure. The peas themselves should be bright green.

Spinach

Greenmarket is the most suitable variety with large fleshy leaves, dark green in colour, and long standing in habit provided the grower is ruthless. The secret with growing annual spinaches such as this is to sow thinly in rows 15 inches apart, and, as soon as the seedlings show a short leaf, single them down to 12 inches between plants. Sowing may be made from March until August in succession.

Harvest when leaves are still young and bright green in colour.

Home Freezing of Vegetables

It has always been assumed that blanching is essential to inactivate the enzymes contained in vegetables and ensure a storage life of a reasonable duration, but recent experiments indicate that for many vegetables blanching is unnecessary.

However, until results are conclusive I think it best to continue the laborious job of blanching for most vegetables as a crop failure in the freezer would be even more disastrous than out in the garden.

Blanch sliced carrots for 3 minutes

Cool as quickly as possible

But for those who wish to do their own experiments, or who are prepared to take a gamble, below is a selected list of storage times for unblanched vegetables. I have kept the list short as I feel that many vegetables do not store well unless blanched, and so those listed are a good risk particularly if the storage life is kept well within that suggested.

Vegetables	Storage life
Broad beans	3 months
Carrots	9 months
Courgettes	6 months
French beans	6 months
Mushrooms	6 months
Peas	6 months
Peppers	6 months
Runner beans	3 months
Spinach	9 months
Sweetcorn	1 month
Tomatoes	6 months

For most other vegetables blanching is necessary. Restrict the poundage to be done each day as it is a time-consuming job.

Each vegetable is prepared according to its type—leafy varieties must be thoroughly washed, root vegetables thinly peeled, podded varieties picked through carefully.

Method for Blanching

Prepare the vegetables according to their needs and sort into uniform portions.

1. Bring 6 pints of water to the boil in a large pan.
2. Place 1 lb of vegetables into a blanching basket and introduce into boiling water.
3. The water must return to the boil within 1 minute. The blanching time is then calculated.
4. After blanching, chill the vegetables under the cold-water tap, using ice-cubes if available. Drain well.
5. Pack into family portion units and place in the freezer immediately.

Vegetables can be frozen free-flow by laying on trays and packing when frozen. However, so long as the vegetables are packed in two- or three-portion units which can be used straight from the freezer, this seems a pointless process.

Blanching times must be adhered to, or otherwise colour change will occur together with loss of crispness.

Use only young tender vegetables for freezing. Older or inferior vegetables should be diced and used for stock, soups or casseroles, or frozen as stewpack.

Pack in family-sized units and seal

Packaging Materials for Vegetables

Heavy-gauge polythene bags are quite suitable for vegetables. Containers can be used if available but they are uneconomical on space if freezing large quantities.

Ensure that all the air is extracted from the bag. Seal and label.

Vegetable	Preparation	Blanching time (in minutes)
Asparagus	Select stalks which are young and tender. Grade by thickness for blanching. Cool, drain and pack in semi-rigid containers.	Thin stalks 2 Thick 4
Aubergines	Peel, slice and pack into small units.	4
Avocados	Successful only in purée form with seasoning and lemon juice.	

Beans—French & runner	Choose beans which snap cleanly and are not stringy. Cut off ends. Slice, cut or leave whole according to taste.	Sliced 1 Cut/whole 3
Beetroot	Use young beetroot not more than 2–3 in. in diameter. Cook thoroughly for 10–13 minutes according to size. Cool, slice, cube or leave whole as preferred	
Broad beans	Pod and grade using only young, tender beans.	3
Broccoli	Choose compact heads. Divide sprigs so that they are 1–2 in. wide, 3 in. long. Remove woody stalks.	3
Brussels sprouts	Select firm, tight sprouts not more than 1 in. in diameter. Trim outer leaves and make a cut in the stem.	3
Cabbage	White cabbage is the most suitable. Cut into wedges or shred. After blanching drain well.	2
Carrots	For small whole carrots, remove tops. After blanching cool under running water when skins may be rubbed off at the same time. Larger carrots should be sliced or diced.	Whole 4 Sliced 2
Cauliflower	Break into florets of even size 1½–2 in. diameter.	3
Celery	Choose crisp, tender celery. Remove outer stalks and cut into 1 in. pieces. Suitable for flavouring savoury dishes. Celery hearts should be trimmed to 3–4 in. and all coarse outer stalks should be removed. Celery preserved by freezing is not suitable for use as a salad vegetable.	Hearts 6–8 Stalks 3

Corn on cob	Choose tender but not overripe corn. Remove husk and silk. Ensure the centres are completely cool before packing. For corn kernels cut the kernels from the cob after blanching.	4–8 according to size
Courgettes	Select young, tender courgettes. Leave whole or cut in 1 in. slices.	Whole 2 Sliced 1
Onions	Peel, slice or leave whole. Double wrap to ensure smell doesn't penetrate into the freezer.	2
Parsnips, turnips, swede	Freeze only young, tender vegetables. Prepare in the usual way and dice, slice or leave whole according to taste.	Whole 4 Diced 2
Peas	Young, tender peas only. Pod and grade. Discard any that are tough or discoloured.	1
Peppers	Remove seeds and pith. Leave whole, quarter or dice—a selection of various types is useful.	3
Potatoes—new	Only small, tender potatoes are suitable. Scrape and cook almost completely in the normal way. Cool, pack, freeze.	
Potatoes—old	Cook until tender, mash and cream. Freeze as croquettes or duchesse potatoes.	
Potatoes—chips	Prepare potatoes in the usual way and deep-fat fry until just turning to a pale brown colour. Drain on absorbent paper, cool and pack.	
Spinach	Pick over leaves carefully—remove discoloured parts and woody stalks. Blanch in small quantities and shake frequently to ensure a thorough penetration.	2
Tomatoes	Freeze whole or in halves. Prepare as purée or juice. Blanching is not necessary.	

Tupperware containers stand up to the wear and tear of freezer use

The selection of vegetables chosen for home freezing will depend on your family requirements, but some are much more worth while than others. For example, there seems little point in freezing vegetables which are available all year round such as cabbage or potatoes. Chips require blanching in oil, which is an expensive and very messy process—they are probably as cheap bought ready frozen. If buying vegetables in season to freeze, consider the wastage involved; on some it is as much as 75 per cent, which again can make the task uneconomical.

Some vegetables do not freeze very well—salad vegetables, for example, are very disappointing and are best always eaten fresh. Other vegetables, such as avocado and aubergine, are not 100 per cent successful, and, as these are expensive even in season, careful consideration should be given before buying. On the other hand, as a way of preserving left-overs, freezing is ideal.

The rule should be to choose selected vegetables according to the family's needs and freeze small amounts, enough only for a few months' supply.

Home Freezing of Fruit
There are various methods of freezing fruit and two considerations must be borne in mind when deciding which to use—firstly the type of fruit to be preserved and secondly the purpose for which the fruit will ultimately be used.

Soft fruits which make their own juice such as raspberries can be packed in dry sugar. Firm-textured fruits with a mild flavour can be improved by freezing in syrup.

All fruit, however, can be frozen down with no additives at all and the result will obviously be a more natural-tasting product. To my mind this is the best method for the majority of fruits. Quantities of the same fruit could be frozen down in several different ways, depending on its ultimate use—a dessert, a sauce, perhaps for jam making at a later date or for the annual spring diet.

Colour coding helps identification of freezer packs

Freshness is the key. Most people eat fruit for its natural taste, and so to ensure these qualities are maintained in freezing, follow one simple rule. Work quickly. For the best results freeze the produce within two hours of picking. Bruised, unripe or overripe fruits are not suitable for freezing, and so be very selective.

Avoid using galvanised iron, copper or brass utensils while preparing fruit. They may taint the flavour or cause discoloration. Fruit can be stored satisfactorily for 8 to 10 months, but to prevent discoloration, which tends to occur with some fruits, plunge them into salted water or lemon juice immediately after peeling or skinning.

Any stones or pips should be removed before freezing. If this is not done, the storage life of the fruit is reduced and also there is a chance that the stone will turn rancid and taint the fruit.

Methods of Freezing

Sugar syrup. This method is ideal for fruits that discolour easily or those with little natural juice. The strength of the syrup can vary depending on individual taste, but the most popular is a solution made up of 12 oz of sugar to 1 pint of water. As the water must be heated to dissolve the sugar, it is advisable to prepare the syrup a day in advance and refrigerate.

Slice large fruits into pieces and immerse the smaller fruits whole in the syrup, allowing half a pint of syrup for every pound of fruit. Crumple a piece of cellophane on top of the fruit to help keep it submerged. Leave half an inch headspace at the top of the container to allow for expansion.

Dry sugar. Prepare the fruit as necessary, taking care not to bruise it. Small fruits should be left whole, others sliced. Work in small quantities so that the fruit is not crushed under its own weight. Lay the fruit on greaseproof paper, sprinkle the sugar over it and leave to stand for a few minutes until the juices begin to flow from the fruit. Shake the paper gently until all the fruit is coated in syrup and pack into the containers, leaving half an inch headspace. Allow about 4 oz of sugar to 1 lb of fruit.

Free-flow. Avoid washing the fruit if possible when freezing free-flow. Valuable juices will be lost and there is no sugar or syrup to substitute. Lay the prepared fruit out on baking sheets or trays. It may be left uncovered for the short space of time

Freezing rhubarb by the free-flow method

required to freeze it. Once frozen, all specks of dirt and foreign bodies will have departed from the fruit, leaving it clean and ready to pack. For fruits such as redcurrants and blackcurrants there is an added advantage—stalks will "freeze off", too, saving hours of painstaking work.

Remove the fruit from the trays and pack in suitable containers—the advantage, of course, being free-flow fruit. Sugar or syrup can be added during cooking or thawing if required.

Fruit purée and juices. Fruit that does not come up to standard can be used to make purée. After preparation pass the fruit through a sieve—it can be cooked first but this is not always necessary. Add sugar according to taste—4 oz to 1 lb of fruit is usual. Purées pack compactly and make economical use of freezer space. They are excellent for baby foods, sauces and delicious toppings to liven up ordinary desserts, particularly ice-cream.

Citrus-fruit juices freeze well so long as care is taken to exclude pith, and sugar may be added to taste. They are excellent breakfast drinks and a useful addition to party punch.

Always have a stock of frozen lemon slices for garnish and drinks.

When frozen, pack in half-pound or pound units

Containers for freezing fruit. Rigid containers are necessary for the syrup and dry-sugar methods. Use plastic containers, Tupperware, Aulafoil bags or heavy-gauge polythene bags within a container (e.g. caster-sugar box).

Heavy-gauge polythene bags can be used to store free-flow fruit, but careful storage is required in the freezer to prevent crushing.

Margarine tubs and yoghurt cartons are ideal if storing purées and fruit juices, but ensure that the lid is airtight. Label clearly.

5. FREEZING MEAT AND FISH

Meat

It would take more than a few paragraphs to convert anyone into a butcher, and so my guidelines on home-freezing meat are assuming the carcass is ready jointed. It is particularly important with meat to pack in the quantities required, or otherwise the tendency is to use more meat and so while the family eats better the savings dwindle.

It is necessary to spend time in preparation before freezing as most of the meat will be cooked straight from the freezer.

Joints

Trim off any excess fat, remove the bone if possible and string the joint if necessary to make it a compact shape. Wipe with a clean cloth. Cover any sharp bones with several thicknesses of foil to prevent outer packaging being punctured. Double-wrap the joint, first in cellophane or foil, then into a polythene bag.

Steaks

Trim off excess fat. To ensure that the steaks are free-flow, wrap each one in cellophane and stack neatly into a polythene bag. Chops and offal should be treated in the same way.

Meat for stews and casseroles

Prepare the meat as for cooking. Remove excess fat and cut into cubes or mince. Wrap in family-size portions and place three or four packs in a polythene bag.

To prevent freezer burn always double-wrap

Packing materials for meat

For the first wrap use vapour-proof cellophane or aluminium foil—both can be re-used if treated carefully. Heavy-gauge polythene bags are most suitable for the second wrap. Expel all the air. Seal and label.

Poultry and game

After being hung and drawn, poultry and game can be frozen down in several ways: whole for roasting or boiling; halved or jointed for grilling, frying or casseroles. Always double-wrap all portions as poultry is particularly prone to freezer burn.

Freeze and store giblets separately as their storage life is only half that of the bird. If several birds are being prepared at one time, store the chicken livers together and use for pâté at a later date.

Fish

Fish *can* be frozen whole, but this should be done only with fish such as trout and salmon, where presentation must be considered. Otherwise all fish should be gutted before freezing. This is time-saving later on and more economical on freezer space. It also prevents discoloration of the flesh.

To freeze fish whole, wash and remove scales by scraping the fish from tail to head with the blunt side of a knife. Gut the fish and remove the fins and eyes. Wash thoroughly and dry.

Both round and flat fish can be skinned before freezing, but this is a laborious job and in most cases unnecessary.

It is a good idea to glaze large fish to give them extra protection in the freezer. Freeze the fish unwrapped and then dip into ice-cold water—a thin film of ice will form over the fish. Return to the freezer. Repeat this process three or four times until an ⅛ inch layer of ice has formed around the fish. Finally wrap the fish in the normal way. The only disadvantage with this method is that the ice must be removed before cooking.

To prepare fillets, cut down the backbone of the fish and remove each fillet carefully, ensuring that all bones are removed. Wrap each fillet individually. Steaks should be cut with a very sharp knife across the backbone; a 1½–2 in. thickness is recommended.

Shellfish

The most satisfactory crustaceans for freezing are prawns and shrimps. They have a relatively short storage life but retain their flavour better than other varieties. They store more satisfactorily if uncooked. Depending on how they are to be served, either leave the shells on or remove and wash in a weak brine solution. Drain, freeze free-flow and pack.

Crabs and lobsters should be cooked before freezing. Place in lightly salted boiling water and simmer for 15 minutes per pound. Drain, remove the edible meat from the shell, and cool. Pack and freeze. The shells can be oiled and stored separately to use for serving.

Packaging materials for fish

Fish should always be double wrapped to protect the delicate flavour. Wrap firstly in cellophane or foil and then pack into polythene bags or containers. Remove air and seal. Label clearly. Fish is susceptible to freezer burn if packed insecurely—this has no harmful effects but renders the fish slightly tough and rather flavourless.

6. HERBS AND THEIR USES

Until a few years ago herbs played only a minor part in British cookery, but lately we seem to be acquiring a new fashionable habit, growing and using herbs not only for entertaining or summer salads but in everyday cooking. I hope the fashion lasts, as it has done much to relieve the boredom of ordinary dishes.

One of the great advantages of herbs is that they are inexpensive to use because you need only small amounts. Their job is to add that subtle something to the dish which delicately enhances the flavour of the main ingredient without overwhelming it. The cheapest and best way of getting the herbs is of course to grow your own and store them in the freezer.

March is the time to start. If possible find a site near the kitchen so that you can pop out at the last minute to snip off a few chives or a sprig of parsley. But if there is no room to devote a complete patch to herbs plant them alongside the garden path, or even mixed with the herbaceous border—some herbs flower beautifully so that they don't look out of place.

Choose a sunny, well-drained position if possible. If you can set aside an area just for herbs, plant them in a geometrical pattern and edge them with a low hedge of rosemary and

No one should be without a herb garden, whether inside or out

lavender, both of which have pleasant mauve flowers.

I suggest starting with the following herbs which, with the exception of parsley, are all perennials: mint, thyme, sage, chives, bay, fennel, oregano and parsley. These can all be sown in boxes in early spring provided they are kept frost-free and planted out later. Alternatively, buy plants and get a quicker start.

Mint

Chances are that you may have the ordinary mint *(Mentha spicata)* in the garden already. It grows easily and because it forms a formidable root system it is best, when growing it with other herbs, to confine the mint by surrounding it with pieces of slate sunk vertically into the ground. Try other mints, too —a couple of sprigs of lemon mint with a slice of cucumber added to a glass of white wine make a most refreshing summer drink. Add soda-water if you want a really long drink. Pennyroyal *(Mentha pulegium)* is a dwarf mint very useful for edging; it has tiny spines of mauve flowers.

Many use mint when boiling potatoes or peas. But try chopping a few leaves and adding them with a knob of butter to the strained potatoes. Also mix a little chopped mint with cream cheese for a dip or sandwiches.

Thyme

There are many species of thyme. Probably the most useful for the cook is garden thyme *(Thymus vulgaris)*. It can be grown easily from seed and forms a tough, shrubby plant useful in edging flower beds. Use fresh in stews and all meat dishes.

Sage

Sage prefers a light soil and a sunny position. The best variety to grow is the grey-leaved one which has a pretty purple flower and forms bushes 2 to 3 feet high. It will grow easily from seed and the bushes should be cut back each year after flowering.

Sage is inclined to get a bit leggy after several seasons, and so make new plants by breaking off roots from the old ones every two or three years. Sage forms the basis of one of our classic stuffings for poultry, especially with onion for roast duck, but it can be used sparingly in other savoury stuffings as well.

Also try serving lettuce and tomato salad with cottage cheese into which are beaten a few teaspoons of freshly chopped sage leaves and a few drops of lemon juice.

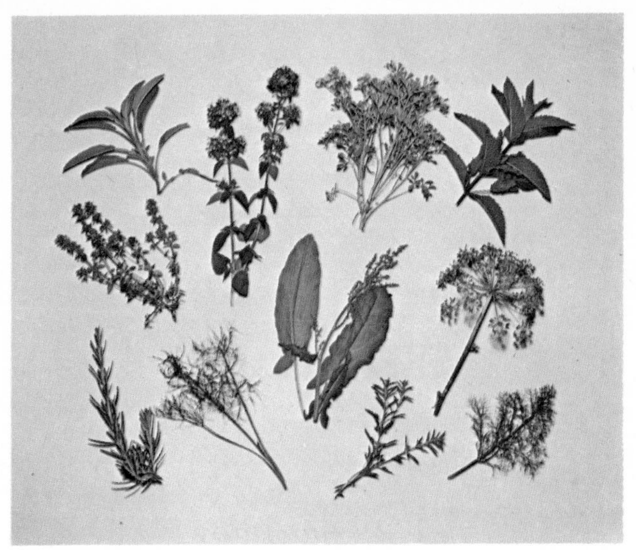

Most common herbs freeze and store well

Chives

These are particularly useful to add a delicate onion flavouring to salads. Chives are from the onion family but without the strong, pungent odour. Seeds should be sown in spring and if the leaves are picked regularly the plants will grow for years. Split up the clumps every two or three years.

Chopped chives are delicious in omelettes—just add a few to the beaten eggs and cook in the usual way.

Bay

It is best to buy a bay tree from the nursery. The tall, shaped ones are expensive, but you need buy only a small bush. Keep it in a large pot or tub which can be moved near to the house for protection when very hard frosts are expected. Do not let the soil dry out completely in summer.

Bay can be used, fresh or dried, in a *bouquet garni*. Tie a leaf together with a sprig of thyme, parsley and marjoram if available, and use the *bouquet* to impart an added flavour to stews and similar dishes. Use powdered bay leaf mixed with other herbs in savoury mince, rissoles, etc., for an interesting flavour.

Fennel
Of the two common types of fennel, Florence fennel grows best in Britain. It can be raised easily from seed and when planted out likes a sunny, well-drained site. It grows into a tall, feathery plant with masses of attractive flowers in late summer and so is ideal for the herbaceous border.

All parts of the fennel plant can be used in cooking. Collect and dry some of the seeds to press on to home-made bread and rolls. The plant has a pleasant, slightly aniseed flavour which goes well with fish. The main stem can be chopped, simmered and used in a white sauce; and the feathery tops can be placed on top of grilled fish for the last minute or two before serving—it gives a distinctive and aromatic flavour which is most pleasing. The feathery tops can be stored in the freezer for several months.

Oregano
This herb is ideal for use with tomato dishes—somehow it complements the tomato flavour beautifully. It can be grown from seed and if necessary the plants can be divided in autumn. The leaves and flowers should be dried together, then rubbed and stored. As well as using with tomato dishes, put a little oregano on steaks before grilling.

Parsley
This is the only herb mentioned which is a biennial, and I find it advisable to grow some from seed each year. It is best sown *in situ* in fairly rich open soil, and as it takes a long time to germinate I sprinkle the seeds with very hot water immediately after sowing.

Parsley is best used fresh and it can be stored for several months in the freezer, but dried parsley can be used in many meat dishes. The fresh parsley is often used as a garnish for fish and steak dishes, and chopped and combined with a white sauce makes an excellent accompaniment to fish dishes.

Rosemary
It grows easily to a bush 2 or 3 feet tall and is an evergreen. It is good with white meats, fish and chicken.

Traditionally herbs have been preserved by drying and storing in airtight jars, but the freezer enables you to store fresh herbs for use during the winter months. It is best to preserve small quantities of each in both ways as each plays a role in cooking. To dry, hang in a cool, dry place for a couple of weeks, strip the leaves and rub through your hands. Store in an airtight jar.

Freezing gives more satisfactory results for many herbs, particularly chives and fennel, as it retains the true subtle flavour sometimes lost when drying. Freeze the herbs separately on a tray; once frozen, store in polythene bags and label. While frozen, run the rolling pin over the polythene bag—the brittle herbs will shatter, saving the laborious job of chopping at a later date.

Don't forget to save a few pieces intact for garnish. Most herbs will store for six months in the freezer, but don't store too many—a little goes a long way.

Aluminium foil containers such as this Alcan pie-dish are particularly suitable for freezing and cooking

7. PREPARED FOODS

Most recipes or home-made concoctions will freeze—some more successfully than others—inevitably there is a certain amount of trial and error. But don't think that the only recipes to use are those found in freezer books, as there is rarely anything special about them—the old family favourites can still be used, perhaps with the odd amendment in places.

Always freeze down in family or individual portions for easy use—it relieves the irritation of trying to separate frozen food and it saves the unnecessary job of re-freezing, which can only be detrimental to the produce.

Pastry

Frozen pastry can be thawed and refrozen to facilitate making pies and other dishes. The filling in many cases will be precooked, but the pie itself will be best cooked from the freezer as required.

If, before the pie is made, the dish is lined with foil, the pie can be removed from the plate or dish after freezing and stored in a polythene bag. This will release the cooking dishes for further use. When required, return the pie to the dish before cooking. Left-over pastry should be made up into a dish, but if this is not possible wrap in foil and freeze to re-use at a later date. Puff and flaky pastries in particular are time-consuming to make and often unsuccessful—to my mind the frozen branded varieties are an essential ingredient in the freezer.

Stews and Casseroles

Always cook two or three times the quantity required. Eat one and freeze two family portion packs for future use.

Use seasoning sparingly as an excess reduces the storage life. Season when re-heating rather than before freezing. This applies particularly to garlic and curry. Wine is fine if used in small quantities but it is best to add it afterwards if large quantities are required. When freezing, bear in mind the shape and how it is to be heated.

A good method of freezing is to line the pan or dish with foil, leaving ample to cover the food. When the food is frozen, remove it from the utensil, wrap in the foil and place in a polythene bag. This not only releases the dish for further use but also ensures that food is frozen in a shape suitable for re-heating. If freezing large quantities, it is perhaps more practical to freeze

A lobster-tail casserole such as this can be easily prepared

in block form and cut into shapes.

Pour the stew into a large container, perhaps a roasting tin, and freeze uncovered for a couple of hours. Before it is completely frozen, cut the stew into brick shapes and wrap in foil or polythene. The advantage of this method is the economical use of freezer space with evenly shaped packages. Foil containers are useful, too, as the stew can be transferred straight from the freezer to the oven. Boil-in-the-bag kits are also now available to the housewife—this is a useful way of doing one-portion servings. However, although convenient, it is not particularly cheap. The best method of packaging will depend on family needs, but it should not be necessary to spend a lot of money on buying special materials.

Bread and Cakes

Bread and cakes, whether bought or home made, freeze and store well so long as they are put into the freezer fresh, preferably on the same day as baking.

Bought produce should be packed carefully in suitable containers or polythene bags and placed in the freezer. Home-made cakes can be frozen cooked or uncooked, although it is easier to store them cooked, but not decorated. Decoration can so easily become damaged in the freezer, and also cream and icing can lose texture and colour during storage. If it is necessary to store a decorated gateau, freeze uncovered; once solid, place in a suitable container.

Bread can be stored at any stage. Unrisen or risen dough can be stored for a few weeks, and this is economical on freezer space. Cooked loaves will store for much longer. Sandwiches store for two to three months and are useful for members of the family requiring a packed lunch each day, or for instant picnic outings. Choose fillings carefully—avoid those which will turn the bread soggy on defrosting.

Breadcrumbs and croûtons will store for a few months. Fresh yeast surprisingly survives very well at freezer temperature for a considerable time.

Liquids and Sauces

Fruit juices, left-over stock, gravy, sauces and soups should be frozen in small quantities in concentrated portions if possible. Reduce the water content by boiling rapidly. This can be re-introduced when thawing or re-heating. When freezing in containers, leave 1 inch headspace to allow the contents to expand.

Ice-cube trays are useful, too—freeze the liquid in portions

Yeast products often taste better after freezing

and store in polythene bags. This is particularly handy for concentrated fruit juices, stock and sauces.

Baby Foods

A great saving can be made by making your own baby foods from family meals. When preparing a recipe, extract baby's portions before adding extra seasoning. When cooked, these portions should be liquidised and packed in one-portion sizes, depending on the age of the recipient.

Ensure that the food is defrosted thoroughly before feeding. Hygiene is obviously of the greatest importance and precautions must be adhered to strictly.

Meat & Poultry Dishes

Stews and Casseroles

Preparation

Prepare according to the recipe and cook until the meat is only just tender, to allow for re-heating. Season sparingly and avoid using too much wine, liqueur, curry or garlic. Do not thicken at this stage.

Freezing

Cool as quickly as possible and freeze in the portion sizes required (don't forget individual portions are extremely useful). Pack in one of the following ways:

1 *Using rigid freezer containers or foil dishes.*
2 *Freeze in block form then cut and pack in bricks.*
3 *Freeze in lined cookware and repack to release utensil.*

Serving

Reheat in a casserole or saucepan from frozen.

1 *Preheat the oven to 400°F, Gas No. 6, and allow at least an hour to warm through. Reduce the temperature and continue until hot enough to serve. Allow approximately 1½ hours for a 4-portion stew.*
2 *Simmer very gently in a covered saucepan until thawed and thoroughly heated. Allow approximately 1 hour for a 4-portion stew.*

Check seasoning and adjust if necessary. Thicken if required and add extra wine, liqueur, curry or garlic at this stage.
For goulash and stroganoff, add yoghurt or cream just before serving.

Pastry Dishes · Flans

Savoury and Fruit Pies

Preparation

Prepare the filling as normal. All savoury fillings should be undercooked slightly to allow for reheating. Season according to taste, but "underseason" rather than "overseason". Fruit fillings can be cooked or uncooked according to preference. Sweeten to taste.

In either case if the filling is cooked prior to putting in the pie, cool and refrigerate first.

Make up the pie or flan using home-made or bought pastry. Use foil dishes or foil-lined cookware.

Freezing

Freeze uncovered. When firm pack in containers or polythene bags. Store carefully to prevent damaging. If they are to be stored for a period of months double wrap the pastry items to prevent drying.

Serving

Place frozen pies in a pre-heated oven of 425°F, Gas No. 7, for 35–45 mins. Extra time may be required for large savoury pies. Reduce the temperature after 45 mins. and cook for a further 15–30 mins. A meat thermometer will help determine if the heat has penetrated through to the middle.

If required, after 15 mins. remove the pie from the oven and glaze with egg and/or milk.

N.B. *Pastry dishes can be frozen already cooked. Prepare as above and cook in the normal way. Pack when cool and freeze. Thaw at room temperature for 2–4 hours. If required, reheat after defrosting in a moderate oven.*

Soups, Sauces and Stock

Preparation

Prepare in the usual way taking care not to overseason. Thickening may be necessary for convenience later, but avoid this if possible in the case of soups and stock.

Freezing

Freeze in rigid containers leaving at least ½ in. headspace. Use ice trays for small quantities.

Serving	Thaw at room temperature for an hour or so before reheating. Stir continuously over a very gentle heat until ready. Adjust seasoning if necessary.

Prepared Fish Dishes, Fish Cakes, Mousse, Kedgeree

Preparation	Many dishes require flaked fish, so obviously the fish must be cooked first, but do not overcook as it becomes dry. Make up the dishes in the normal way—a moist mixture is preferable. For other dishes, such as fish in sauce or wine, only half cook before freezing. Season sparingly and do not thicken.
Freezing	Freeze in rigid containers or in foil-lined dishes and repack when frozen. Double wrapping helps to preserve the delicate flavour of many fish dishes.
Serving	For mousse, thaw in the refrigerator overnight. Other fish dishes such as kedgeree should be reheated in a moderate oven for 30–40 mins. from frozen. Fish cakes can be fried or grilled. Prepared fish dishes in sauce should be thawed at room temperature for 3–4 hours, then adjust seasoning and return to the oven to continue cooking. Although not so convenient this method preserves the flavour of the fish and sauce and prevents overcooking.

Cakes

Preparation	Cakes are best frozen cooked but not decorated. Bake in the normal way according to a favourite recipe. Take care not to "overflavour" with spices or essences. If decorating is to be done at this stage to save

time later, avoid using jam and cream. Icings and butter creams freeze quite well in reasonable amounts.

Freezing

Wrap each cake or portion individually in cellophane, wrap and pack carefully in rigid containers.

Decorated cakes should be frozen unwrapped and packed when firm.

Serving

Thaw at room temperature

Large cakes 3–4 hours
Small cakes 1–2 hours

Iced cakes and fruit cakes could take up to 6 hours depending on their density. Unwrap decorated cakes before thawing to prevent the wrap sticking. Plain cakes should be left wrapped.

N.B. *Uncooked mixtures can be frozen but require complete thawing before baking to get the best results.*

All baked yeast mixtures freeze well. Thaw at room temperature for 2–4 hours.

Bread and Bread Dough

Preparation

Bread can be frozen at various stages—unrisen dough, risen dough or baked.

In all cases prepare according to the normal recipe but if intending to freeze bread dough use a little more yeast in the mixture.

Enriched yeast mixtures such as croissant and teabreads should be baked before freezing for the best results.

Freezing

Baked products should be wrapped in foil and then sealed in a polythene bag.

Place unbaked bread dough in a lightly greased polythene bag leaving headspace for expansion.